What Does Life Do?

Understanding Life as a Process

TOM KENNEDY

Kennedy Science Productions, LLC
Albuquerque, NM

Printed in United States of America

First Printing, 2016
ISBN: 1-945331-00-3
ISBN-13: 978-1-945331-00-8

Kennedy Science Productions, LLC
www.ksciproductions.com

FOR MY WIFE

Jen,

My best friend and life-long companion,
I could never have done this without you.
You're super awesome!
I love you so much!
Thanks for making sure I enjoy every moment of life.
Especially with pizza and beer.

CONTENTS

ACKNOWLEDGMENTS

I could have never done this on my own. Most importantly, I thank my family including my wife for her unwavering support, my parents, and my brother and sister-in-Law have all been instrumental in supporting me at every step in life. John Rogers for reviewing the first draft and providing valuable feed back. My friends, Ian Latella, Eric Schaad, Mason Ryan, Jay Fuller, Steve Poe, Blair Wolf, Michael Collins, Bart Kicklighter, Hudson Cheng, and others of which I have had numerous adventures and late night discussions with that helped me grow as a person and a scientist. And lastly, my PhD advisor Tom Turner who has been both a mentor and a friend.

Introduction

It would be safe to say that most of us intuitively know that a living organism is different from a rock. But, at our most basic level, a human and rock are made of the same fundamental building blocks of matter as everything else in our universe, from the rocks that form our planet, to the air we breathe, to the stars in the night sky. The question becomes, if we are all made out of the same fundamental building blocks, what really separates life from every-day objects? How is a living organism different from a rock.

For me, the key to understanding what makes life unique lies in how we view life. Perhaps, a good way to think about life begins by asking *"what does life do?"*, rather than asking *"what is life?"* The question, "What is life?" implies that life is a noun or a list of characteristics to be memorized in an introductory biology course. I can recall finishing my PhD in biology at the University of New Mexico when someone asked me a seemingly simple question, "what is life?" At first, I began to recite a list of characteristics describing life similar to what you often find in the first chapter of popular biology text books. Shortly after reciting two or three of the characteristics, I couldn't remember the others.

Admittedly, it was embarrassing that I couldn't adequately define life based on text book knowledge. However, it was at that moment I began to realize life was something more than a list of characteristics; life is an action, it does something. Life is about processes, and it is those processes that makes life different from non-living objects. I also began to realize that the processes of life connect us to the universe through energy flows and nutrient cycling.

After my first five years of teaching, I decided to write a book about biology with a focus on life as a process. While it's impossible to cover the entirety of biology in single book, I cover life's origins, evolution, diversity, connections to the Earth, and end with my speculations on the future of life and humanity. And of course, I also added many of my favorite biological stories that I find most interesting.

I set out with three major goals for this book. The first is make biology interesting and exciting. Second is to create a broad view of life by incorporating all the fields of science into biology. And third is to illustrate how science works, that it is an iterative process to understand the natural world.

First Goal: To make biology interesting. I love biology, and it breaks my heart whenever a student tells me that they hated their biology class. Perhaps the worst part is that I understand their sentiment. I too have suffered through memorizing facts without any context of their meaning, or dissected some preserved cat without understanding or appreciating what I was looking at, but merely as an exercise to memorize some body parts. I too have memorized the steps of a dividing cell or the parts of a flower just because it was required of me. Science is not a collection of facts, and neither is biology. In this book, I do my best to make biology interesting and not simply rehash facts. I know that science is not for everyone, but perhaps you will find something interesting within these pages.

Second Goal: To create a broad view of life by showing what all living organisms must do in order to be alive, regardless of how small, large, or bizarre they appear. To fully appreciate life, we need to know a little astronomy, physics, chemistry, geology, and climatology, basically a little of all the sciences.

Life is connected to the universe. It may be difficult to believe, but the origins of life were made possible by astronomical events starting with the Big Bang, a single moment in time when all the energy and matter in the universe were created. Later, exploding stars billions of years ago created the heavier elements that would form our Earth and the building blocks of life. Today, we still depend on our sun for a constant supply of energy to fuel life.

Life is connected to the Earth. In an ancient ocean sometime 3.7-4 billion years ago, the building blocks of life began to assemble into more complex molecules until life emerged as an extension of geological processes. Slowly, life emerged as a process that takes in energy from the environment to create order from chaos by assembling increasingly complex molecules. Life is a system out of equilibrium with its environment, where death is a process returning to equilibrium. At some point, those molecules acquired the ability to store information and replicate themselves to perpetuate their processes.

Since its simple beginnings, life has evolved to become a major force shaping the Earth's surface. We can thank our oxygen-rich atmosphere to photosynthetic organisms that first evolved billions of years ago. Oxygen reacted with iron dissolved in the early oceans forming the banded iron formations we mine today to make steel. As iron was slowly removed, the oceans assumed their blue color we see today. These early photosynthetic organisms also paved the way for the evolution of multicellular life that would come to dominate the Earth's surface.

The evolution of life is made possible because life reproduces by passing on information from one generation to the next, ensuring the continuity of life. In each generation, small errors accrue as mutations when genetic information isn't perfectly copied. It's these errors that provide the variation among organisms vital for the evolutionary forces of natural selection to act upon. Since its beginnings, life has evolved as genes have been passed down for countless generations, slowly changing over time. In fact, you are the end result of an unbroken lineage going back about 3.7 billion years, a profound statement that I will state again!

No living organism is an island, existing alone or isolated. All life must interact with its environment to acquire nutrients and energy to create order. Energy and molecules flow through our bodies like water flows through a river, connecting every living organism to the Earth. Unlike energy which cannot be recycled, the elements forming molecules are continuously recycled, made available by special types of living organisms. Every carbon atom currently in you was once in a molecule of carbon dioxide in the atmosphere until it was fixed by a plant into a sugar by harnessing a practically unlimited source of energy from the sun.

Third goal: I want to show the nature and limitations of science and how we use it to understand the natural world. Perhaps the most important point is that science is really a way of thinking, it is part of who we are. Science begins by making observations and asking questions, which requires a curious mind. Without curiosity and intellectual pursuits, there would be no science. Without science, humans would still be hunter gatherers, living in caves and subject to the vagaries of climate. There would be no cities, civilizations, cars, internet, cell phones, Facebook, or art and literature.

If science is about our curiosity and need to understand the world around us, then science begins with observations and asking questions. But, it doesn't stop there, we must follow up by seeking the answers to our questions through experimental tests, or by making additional observations. Sometimes, we get the answer right the first time, other times we are wrong and we must try again. But keep in mind, science is an iterative process and self-correcting. Almost every field in science has had their paradigms tested and overturned. Even the most beautiful theories can be slayed by a single observational experiment. But, over time, science progresses and our understanding of the world continually grows.

One of the most important aspects of science is that it is limited to phenomena that can be potentially observed, measured, or verified by experimentation, which I will define here as the natural world. Most people not trained as a scientist hold several deep-rooted misconceptions regarding science. One of the most common beliefs about science is that it is open to any possible explanation. Luckily, this is not true or we would have to entertain any idea regardless of how absurd it is.

Could you imagine having to seriously consider the hypothesis that the Earth is carried on the back of a giant, invisible turtle swimming through the cosmos. Sometimes, it's difficult to determine the answers to our questions, like what caused the Big Bang, or why did it take 3 billion years for animals to evolve? But, that's OK, we never stop trying to figure out how things work.

Over time, scientists have developed scientific laws and theories to explain how the natural world works. Scientific theories are broad, powerful explanations for a set of observations and provide a framework for interpreting new results and observations. For biology, the central paradigm is Darwin's theory of evolution by natural selection. It explains the unity and diversity of life by making the prediction that all life descended from common ancestry and we share traits of those ancestors.

Science and critical thinking can be thought of as two sides of the same coin. As we use the process of science to understand the world, we should use scientific information to help us make informed decisions in our everyday lives. Critical thinking is not limited to science, we should always strive to use the best facts and information available to inform our opinions and make decisions. Unfortunately, the findings of science often become politicized, especially when they go against preconceived notions or anecdotal knowledge. Regardless of what someone believes, scientific findings explain how the universe works and are not political statements.

Humans are newcomers to the Earth. In a very short time, we have drastically altered the planet like no other species before us. It is only through science and critical thinking that we can understand the true nature of the destruction we have brought to the Earth. Science can also provide solutions to solve the biggest challenges to humanity including population growth and climate change. To deny science is to deny reality.

Chapter 1

Origins of the Universe

and the Earth

It is far better to grasp the universe as it really is than to
persist in delusion,
However satisfying and reassuring.

Carl Sagan

Introduction

When I'm hiking on the west side of the Sandia Mountains near Albuquerque, NM, I can't help but think that the mountain range is very young for mountains, about 10 million years old. However, the granite forming the mountain is approximately 1.5 billion years old; older than the dinosaurs, even older than the first animals. The granite forming the Sandia Mountains is so old it was formed when all life existed merely as single-celled organisms. If you were to go back in time to the formation of the Sandia granite you would not recognize the Earth, other than it was mostly covered in water.

In some ways, we are similar the granite forming the mountains. Rocks are not living, but they are made of many of the same elements as you and me. Those elements we share with the rocks forming our planet are ancient, older than the Earth itself. They were formed inside stars that exploded billions of years ago spewing their insides to the universe. Even the elements are made of more fundamental building blocks, just three basic subatomic particles, protons, neutrons, and electrons.

For the first time in the history of our species, we are fortunate enough to live in a time when we have the technology to understand the nature of the universe and explain our origins without invoking super-natural explanations. To accompany our technological advances allowing us to observe tiny cells or distant galaxies, we have also developed a methodology, commonly known as the scientific method that allows us to understand the natural world. Importantly, we should all remember that science is not good or bad, conservative or liberal, it's a process of understanding the natural world. It's also important to know that contrary to popular beliefs, science is limited in its scope to understanding the natural world.

The Nature and Limitations of Science

At its root, science means the desire to know. In modern times, science has evolved into a process that allows us to understand the natural world. Science begins by making observations and asking questions. Sometime between elementary and high school, most people stop observing the natural world around them, let alone ask questions about it. By the time most of us are adults, we simply take the natural world for granted.

For those of us who live in Albuquerque, we have a large mountain east of town, abruptly rising 4,000 feet above the city, and the Rio Grande River flows through the middle of town. One thing I've noticed as I've taught thousands of students, is that few have ever asked the simplest questions about our world around us: what caused the Sandia Mountains to rise east of town, how old are they, why are there fossils of ancient marine life on top of the mountain, why are there volcanoes west of town, or how is the ecosystem in the mountains changing due to climate change? These are the types of questions that scientists may spend their careers attempting to answer.

It is through the efforts of thousands of scientists that scientific knowledge has accumulated over time through an iterative and self-correcting process. Hypotheses, which are proposed explanations for a set of observations, are tested, and either kept or discarded based on the data collected. Even the most cherished hypothesis can be quickly proven false and discarded with a single observational fact.

Although science is a very broad subject of inquiry, it is limited to understanding the natural world. When I use the term, "natural world", it has a specific meaning with major implications for the scope of scientific endeavors. The natural world is limited to phenomena that we can potentially observe, measure, and test for. Using these criteria actually places limits on science, which is a good thing. If not, we'd have to entertain any idea to explain our world, no matter how implausible or absurd it is. If there is no conceivable way to make an observation or test for a phenomenon, then it's outside the realm of science and may even be pseudoscience. Pseudoscience makes claims about the world that "sound or appear scientific", but does not actually follow the conventions of science. Examples of pseudoscience include intelligent design, or aliens built the ancient world.

To illustrate a great example of pseudoscience, imagine a scenario where someone came up with the notion that the Earth rests on the back of a giant, invisible sea turtle swimming through the cosmos. You can't test for the turtle's existence, and you can't disprove or prove it's there. To state it bluntly, those kind of fanciful ideas are beyond the scope of science because they are not observable, you can't test for them, and you can't potentially disprove it. It's a good thing that science is limited, that way scientists aren't preoccupied trying to disprove an unlimited stream of untestable ideas. I'm totally fine with that, it gives me comfort that scientists never stop observing, asking questions, and testing their hypothesis to learn about our world, no matter how difficult the question. Because of the continuous efforts of scientists following the conventions of science, our knowledge and understanding of the universe grows continuously and we don't have to entertain supernatural explanations for the world.

In addition to pseudoscience, there are other misconceptions regarding the nature of science. Much of this stems from the way science is often taught in school: as a collection of facts to be memorized, or hypotheses are merely educated guesses. It is true that scientific studies do accumulate facts, but more importantly, science is a process and a way of thinking about the natural world. Hypotheses are borne out of our observations and prior knowledge, they are predictions or proposed explanations, not guesses.

Science is not another belief system. It is not based on unverifiable stories lacking in actual support. Science is based on evidence that we can verify through observations or experimentation. As we improve our understanding of the natural world, science is able to make additional predictions based on our hypotheses, theories and laws. If they fail to accurately predict outcomes, they are modified or scrapped. Throughout this book, you will notice that the scientific theories I present, including the Big Bang theory, origins of life, evolution by natural selection, or global climate change, all make certain predictions that have been repeatedly verified based on scientific evidence.

Biology is a scientific discipline focused on studying life. While in college, I took a chemistry class and the text book was called, "Chemistry, the Central Science". If chemistry is the central science then biology is the apex of science, built on a foundation of chemistry and physics. To fully understand and appreciate the scope of biology it also requires some knowledge of astronomy, geology, and atmospheric science! To illustrate the process of science and how biology incorporates every field of science, I begin with the Big Bang Theory. It may seem odd to begin a book about life with a discussion on astronomy, but for me, it's a logical starting point because it sets the stage for understanding where all the matter and energy we use every day came from.

The Universe Began with a Bang!

Before the 1920s, it was generally assumed among scientist that the universe was ageless, it had been around forever and was mostly static or unchanging. Even Einstein attempted to modify his General Theory of Relativity to accommodate the prevailing theory of his time that the universe was static, unchanging, and eternal. Years later, he called it the biggest blunder of his life. The notion of an unchanging universe began to crack starting in the 1920s when several large telescopes were built, including the observatory at Mt. Palomar, CA. Using this large telescope, evidence indicating a finite age to the universe began to emerge with numerous observations made by astronomers, including Edwin Hubble, the same astronomer that the famed Hubble Telescope is named after.

When these large telescopes were built, very little was known about the nature of the universe, including its age. At the time there wasn't much information, so there really wasn't any reason to believe it had a beginning, an end, or was constantly changing. Progress was made to understanding the size and age of the universe when Hubble began making a series of observations on faint "fuzzy-like" objects known as galaxies. It's hard to believe that at the time, it wasn't even clear to astronomers what a galaxy was. But, after making numerous observations over several years, galaxies were discovered to be very large clusters of stars that were outside of our own galaxy. However, it still wasn't clear how large or old the universe really was.

When making observations of galaxies, Hubble noticed that their light was often shifted to the red end of the light spectrum, we call it a redshift. Because light moves in waves, the wavelength determines the color of light; longer wavelengths would appear red and shorter wavelengths would appear blue. Hubble also showed that the redshift of galaxies was not random, it increased with distance, the further a galaxy was from us, the more reddish it would appear.

The implication of the redshift of galaxies is that they are moving away from us. We know this because the effect of movement of an object on wavelength is known as the Doppler Effect, a phenomenon you have observed every time you have stood on a roadside and listened to a car pass you by. As the car approaches, the sound is high pitched because the sound waves are compressed. Once the car passes you, the sound is lower pitched because the sound waves are elongated. Because light travels in waves similar to sound, it also experiences the Doppler Effect. Light coming from an object moving away from us would appear reddish because its waves were being stretched.

Based on this simple, yet ubiquitous observation, astronomers realized that almost every single galaxy is moving away from us because their light waves are red-shifted. A logical conclusion from these observations is that the universe is far from static and unchanging, it is actually expanding and growing larger! If the universe is currently expanding going forward in time, then if you go back in time, the universe is contracting. As you go back in time, the universe would continually become smaller. Go back far enough, there would be a time when the universe didn't exist, thus placing a finite age on the universe.

Can we determine the age and size of the universe? It turns out we can by making many careful observations of the redshift of galaxies scattered throughout the universe. In the last 20 years, astronomers calculated the universe's age to be approximately 13.8 billion years old. The best known and well supported scientific theory explaining the origins of the universe is the "Big Bang Theory", which predicts the universe began with a cataclysmic explosion creating all the matter and energy in the universe in single instant in time. The idea that the universe began with a cataclysmic explosion was so ground breaking that the British astronomer Sir Fred Hoyle referred to the hypothesis as "the Big Bang" in an attempt to ridicule the theory. In an unlikely twist of fate, the name stuck.

The use of the term 'theory' in science has specific meaning. A scientific theory is a broad, powerful explanation for a set of observations. In this case, the Big Bang Theory explains the observations of an expanding universe and predicts a finite age of the universe. Most good theories, including the Big Bang, also make additional testable predictions. If the universe began with the Big Bang, then there should be additional scientific evidence to support this theory. For example, if the universe began with a cataclysmic explosion, then shouldn't some of that residual energy still be present as cosmic background radiation?

The discovery of this cosmic background radiation was accidently made by scientist from the Bell Laboratories in the 1960s, when a state-of-the art radio telescope detected "background-noise" from all parts of the sky. If you've ever heard static on the radio, that's similar to the cosmic background radiation. At the time, the scientists had no idea what was causing the "hum" from the radio telescope. They checked all the wiring, removed bird and rodent nests, and even had to rule out the implausible idea that the former Soviet Union was beaming radio waves at us. After several frustrating years and numerous attempts to "fix" the radio telescope, it became apparent that the noise was the residual energy of the Big Bang. They had discovered the cosmic background radiation predicted by the Big Bang theory, thus lending additional support to the theory.

Hubble's observation of the redshifts of galaxies is a scientific observation, which can be taken as a fact. A fact is something that exists or actually happens, much like it's a fact the sun sets in the west and rises in the east. To a scientist, observational facts lead to more questions, such as why are the galaxies moving away from us. To answer their questions, scientists propose hypotheses, which are proposed explanations for their observations. To be a scientific hypothesis, it should be potentially testable and falsifiable. I use the word potentially because it may not be possible to test some hypothesis because it costs too much money or we don't yet have the technology, or it may not be ethical.

Originally, the expanding universe started as a hypothesis proposed by George Lamaitre in 1927 to explain the redshift of galaxies and why they were moving away from us. Like other hypothesis, it was testable, potentially falsifiable, and better yet, it had the ability to make predictions about the universe that were later verified by observations, including the cosmic background radiation. Even before Hubble determined that the galaxies were moving away from us, Einstein's general theory of relativity, which explains how gravity works, predicted an expanding universe. For nearly 100 years, scientific observations and additional testing have supported the Big Bang theory predicting an expanding universe with a finite age. The accumulation of additional evidence has helped the Big Bang theory grow to become the most well-supported scientific theory to explain the origins of the universe. Since its start, it has yet to be disproved by a single observation.

The development of the Big Bang theory also illustrates the process of science: observations lead to questions that are turned into testable hypotheses. With further study, hypotheses become well supported, grow in their scope, and eventually accepted as theories. Making observations, formulating hypotheses, testing hypotheses, and using data to refine our explanations is the hallmark of the scientific method.

Although the process of science is a broad, powerful tool for understanding the natural world, it does have limitations. Some questions, such as, "what existed before the Big Bang" may be very difficult to address simply due to a lack of data because the information has been lost, or may not even exist. Despite these hurdles, scientists continue their quest to answer difficult questions by putting forth and testing many hypotheses. We discard the ones that don't work and keep the ones that are well supported. Just because a question is difficult, does not mean it is not worth asking. If you never try to ask the difficult questions, they won't ever get answered. It is the tenacity of scientists to keep asking questions, no matter how hard, that keeps pushing the boundaries of knowledge.

What is Matter and Energy?

Why is the Big Bang important to biology? To put it simply, the Big Bang was the creation of all the matter and energy in the universe. We define matter as something that occupies space and has mass. This is known by physicist as the Pauli Exclusion Principle that states: No two particles of matter can occupy the same place at the same time. There are many particles of matter, but the ones that are most important to our current understanding of biology are three subatomic particles known as protons, neutrons, and electrons. It is these three particles that interact to form atoms.

Atoms are the smallest particles of matter that have the properties of an element, such as oxygen, gold, or calcium. The atomic structure of an atom consists of nucleus with positively charged protons and neutrall charged neutrons, both roughly equal in mass. The number of protons in the nucleus determines the atomic number of an element. Change the number of protons in the nucleus and you have a different element. The atomic mass of an element is determined by the number of protons and neutrons in the nucleus. In a cloud surrounding the nucleus are negatively charged electrons, which are a thousand times smaller in mass than protons. In any uncharged atom, the number of electrons always equals the number of protons. The simplest element is Hydrogen with one proton and one electron and accounts for 75% of the elements in known the universe.

The Big Bang also created all the energy in the universe. Energy is a property of objects, and is often defined as the ability to do work. Broadly speaking, there are two types of energy, potential and kinetic energy. Kinetic energy is the energy of motion, the faster an object is moving the more energy it has. Imagine, you're walking along texting on your cell phone and you bump into a wall, it's unlikely to cause any harm. But, if you were running really fast into a wall, the outcome could be grim because you have increased your kinetic energy. Potential energy is the stored energy of an object, which is often dependent on its relative position to other objects. We know this based on every day observations, drop a glass from a few feet above the floor, it is likely to break due to its relative position to the floor; tip it over a table and it doesn't break. There are numerous ways to store energy, including, water behind a dam, charging a battery, or energy stored in chemical bonds that form molecules.

In physics, the laws of thermodynamics govern how energy behaves. All living organisms are subject to the laws of thermodynamics, there are no exceptions. The first law of thermodynamics tells us that one form of energy can be converted into another form. For example, when you start a fire, potential energy in the wood is converted to kinetic energy that we see as light and feel as heat. The first law of thermodynamics also tells us that energy cannot be created nor destroyed, rather remains constant! This is awesome, because the first law would imply that we should never run out of energy.

Unfortunately, there is the second law of thermodynamics, which states that every time energy is used, meaning it is transferred or transformed, the total entropy of the universe increases. Entropy is a measure of disorder or randomness. If you were to take your clothes out of the dryer and throw them on the bed, then that would be high entropy, or disorder. If you were to hang your clothes in a closet and organize them by color, then you expended work to lower entropy and create order. Unfortunately, despite our best efforts, the universe is grinding to a state of maximum entropy because high entropy is stable. Just think, if you quit cleaning your house, it would reach a state of equilibrium where everything was just randomly distributed in each room. It requires energy to do work to keep a clean and organized house. The continual increase in entropy every time energy is used is the reason that energy cannot be recycled and perpetual motion machines cannot exist. It also has implications for life in that it must have a continual supply of energy. I'll discuss energy further in Chapter 3.

Scientific laws, such as the laws of thermodynamics, are used by scientists to predict a certain outcome under the same conditions. The laws of thermodynamics govern energy transformation and predict that energy can't be created or destroyed, and entropy increases every time energy is used. Unlike theories, scientific laws do not explain why or how something works.

To further understand the difference between laws and theories, I can use gravity as an example. Isaac Newton discovered the law of gravity in the 1680s when he determined that an object will fall to the Earth at a specific rate each time it is dropped. Newton's law of gravity makes a specific prediction regarding the rate an object will fall to the Earth. However, Newton's law of gravity does not explain how gravity works, or what gravity is. It took another two centuries and an intellectual leap for the nature of gravity to be explained. In 1915, Einstein published his General Theory of Relativity explaining how gravity works by predicting that the mass of large objects causes a curvature in space-time. General Relativity was a revolutionary theory years ahead of its time, predicting an expanding universe and black holes before there was any observational evidence to support them. Since its publication over 100 years ago, the General Theory of Relativity has been repeatedly verified through observations and experimental tests and remains our best explanation for how gravity works.

The Origin of the Elements

If you have ever walked into a science classroom, you may have noticed a large chart on the wall known as the periodic table of elements. In all, there are about 92 naturally occurring elements arranged in order of their atomic number. At the top of the chart are hydrogen and helium, the simplest and most abundant elements in the universe and were formed shortly after the Big Bang. Where did the other 90 elements come from? The answer to their origins lies in stellar processes that we can observe today. Based on our best evidence, it took about 200 million years after the Big Bang for the first stars to form.

Stars form when gas clouds of hydrogen collapse under the force of gravity. This causes their cores to heat up as they are subject to immense pressure. Eventually, a point is reached at about 14 million degrees Celsius when the repulsive force of the positively charged protons is overcome and the protons stick together to form the element Helium. Fusing protons into new elements is called nuclear fusion, and it also produces another subatomic particle called a neutron. Nuclear fusion releases an enormous amount of energy, enough to counteract the force of gravity and stop the contraction of the gas cloud as the size of the star stabilizes. We see some of this energy as the light being emitted from the sun. Once nuclear fusion begins, a star like our sun, is born. At the center of our sun, the temperature is about 15 million degrees Celsius and 250 billion times the pressure at sea level!

Stars don't last forever, eventually they run out of hydrogen and nuclear fusion stops. At this point, the star will begin to collapse from the force of its own gravity, further heating its core causing the fusion of helium atoms into other elements. Through a series of repeated contractions of stars at the end of their lives, additional elements such as carbon, nitrogen, oxygen, sodium, chlorine, calcium, and potassium are created over millions of years, or tens of thousands of years for very large stars. Eventually, some of these massive stars will begin to produce Iron-56. Once this happens, the energy released from nuclear fusion is no longer sufficient to prevent the rapid collapse of the star from its own gravity. The collapse of these large stars is so rapid, nearly 70,000 km/s, that the outer layers hit the core and then rebound causing an enormous explosion called a supernova.

Supernova explosions are among the most impressive astronomical events in the universe, they can release as much energy in a few moments as our sun will release in millions of years! When stars explode, the elements formed from nuclear fusion are scattered into space. What I find most interesting is that the force of the explosion is powerful enough to create additional heavy elements beyond iron including gold, silver, mercury, lead, and uranium. The gold or silver we wear for decoration was created during an enormous stellar explosion! Life on Earth, including us, is comprised of 96% hydrogen, carbon, nitrogen, and oxygen. With the exception of hydrogen, all the elements in our bodies were created in stars and spread by supernova explosions that took place more than 5 billion years ago. We are stardust, the remains of ancient stars that ended their lives in spectacular explosions billions of years ago.

Origin and Age of the Earth

The origin of our solar system began approximately 5 billion years ago when a large star reached the end of its life and exploded as a supernova scattering its contents into local space. Our solar system hadn't formed yet, it was a nebulae, nothing but a cloud of gas and dust. A shock wave may have tiggered the nebula to collapse forming the sun and accompanying planetary system. Unlike earlier star systems in the universe, our planetary system is a second or third generation solar system and contained many more elements created from the remnants of earlier supernova explosions. As the center of the nebula began to collapse under the force of gravity, it picked up most of the matter in the nebulae eventually forming the sun. Smaller fragments containing rocks and other metals also began to grow and accrete into smaller bodies eventually forming the planets.

Although the sun accounts for 99% of the mass of the solar system, 8 planets eventually formed along with numerous smaller bodies, such as comets, asteroids, and planetesimals with Pluto being the most famous one. The inner four planets are small and rocky, three of which have atmospheres. The four outer planets are much larger and are referred to as the gas giants. In recent years, astronomers have found some evidence that another large planet may have formed along with the eight planets we see today, but was ejected out of the solar system 4 billion years ago.

The age of the Earth and the solar system has been determined to be approximately 4.6 billion years old by using radioisotope dating. To understand radioisotope dating, we need a little knowledge regarding atomic structure. All the elements are comprised of neutrons, protons and electrons. If the number of protons changes, you have a new element. However, an element can vary in the number of neutrons in the nucleus without changing its chemical properties. Different numbers of neutrons form an isotope, but the atomic number remains the same. However, the atomic mass of an element changes with the number of neutrons.

Some configurations of protons and neutrons are very stable, forming stable isotopes. For example, carbon atoms always contain 6 protons, and the vast majority of carbon atoms contain 6 neutrons for an atomic mass of 12. Rarely, a carbon atom will contain 7 neutrons with an atomic mass of 13. Both Carbon-12 and Carbon-13 are stable isotopes because they don't decay into other elements. In fact, the stable isotopes, such as carbon-12 are so stable we are not even sure how long they could last. Estimates place the lifespan of atoms somewhere between 10^{25} to 10^{34} years, an unimaginably long time considering the universe is only about 13.7×10^9 years old. To put that into perspective, an atom could last 10 trillion times longer than the current age of the universe.

Some isotopes have too many or too few neutrons making them unstable. They are known as radioisotopes because they emit subatomic particles at a specific rate as they change into other elements. This decay rate is known as the half-life of an isotope and does not change over time. Each radioisotope has a unique half life; for exampe, the half-life of Carbon-14 is about 5,730 years. Here's how it works; suppose you have a pound of a radioactive substance like potassium-40. It has a half-life of 1.25 billion years, so after 1.25 billion years you would have 0.5 pounds, after 2.5 billion years, you would have 0.25 pounds. Every time 1.25 billion years pass by, you would have half the amount you started with.

Every element has radioactive isotopes that decay at specific rates, and the half-lives of naturally occurring isotopes have an enormous range from a few minutes to billions of years as in the case of Uranium-238 with a half-life of approximately 4.5 billion years. Scientist interested in the age of the solar system or the Earth use radioactive isotopes found in meteorites, ancient rocks on Earth, and moon rocks returned by the Apollo Missions. Each piece of evidence supports a similar age for the Earth and solar system to be approximately 4.6 billion years old.

Science skeptics have often asked how do we know that radioactive isotopes have always decayed at the same rate? This is a good question and took years to answer. Scientists have been studying the nature of atoms for over a hundred years and have developed theories to explain how radioactive decay works. Based on numerous observations and experiments, and our current knowledge of atoms, we know that radioactive decay is based on the statistical decay rate of a population of radioisotopes and doesn't change over time. Additionally, the conditions under which most radioistopes formed don't exist on the Earth, so there are no forces present that would alter their decay rate.

Another reason why we have good reason to believe that the half-lives of radioisotopes do not change is based on the principle of uniformitarianism. Uniformitarianism is the assumption that the same natural laws that apply on Earth are the same throughout the universe, at any point in the past, or at any point in the future. In the case of radioactive decay, this means that the rate Uranium-238 decays into lead, its half-life of 4.5 billion years, is the same today, the same in the past, and the same anywhere else in the universe.

Uniformitarianism is a concept borrowed from geologists in the early 1800s wishing to understand how current geological processes can inform us about the past. Over time, the assumption has been repeatedly supported by numerous observations and experimental verification. To date, there are no scientific reasons, observations, or experiments that have refuted uniformitarianism, this basic scientific principle has withstood the test of time and scientific verification. To state it bluntly, we don't have any reason not to believe it. Uniformitarianism is quite useful to understanding the universe because it would be almost impossible to understand the natural world if the rules were constantly changing!

A *Very* Brief Summary of Earth's History

The Earth is ancient, not as old as the universe, but still far older than most of us can easily comprehend. To organize Earth's history, geologists have created geological time units based on major changes in geological processes or based on major events in the history of life. The largest unit of time is the eon, which can last for billions of years. Eons are subdivided into eras that span tens of millions of years, eras are subdivided into periods, which are further subdivided into epochs. Below, I provide a very brief description for some of these time periods.

Most of us have a difficult time comprehending extremely large numbers such as the age of the universe or the age of the Earth. To help us understand the immense age of the Earth, let's take an imaginary road trip back in time. Imagine that one millimeter, the width of the period at the end of this sentence, represents a year in time. Therefore, someone who has lived 25 years would represent 25 millimeters in length, or 2.5 centimeters, which is about an inch. Not a very long road trip! The average life span in the US is about 85 years, which would be 85 millimeters or about 3.3 inches. To go back a thousand years, you would move one meter, or a little over three feet. To go back to the late Pleistocene Epoch 22,000 years ago when glaciers dominated the northern hemisphere, we would only have to travel 22 meters, or about 72 ft.

To travel back in time to the origins of the first modern humans 200,000 years ago, we would walk a mere 200 meters, a little longer than two football fields. If we were to travel back in time to witness the extinction of dinosaurs 65.5 million years ago, we would have to travel 65.5 kilometers or about 39.3 miles. If were to travel back in time to observe the origins of the first animals sometime between 550 and 600 million years ago, we would have to travel about 550-600 km (330-360 miles) a decent 6-hour road trip on an interstate highway. To go back to the beginning of the Earth we would have to travel 4600 km (2700 miles!), a trip this long could take you from Albuquerque, NM to New Brunswick Canada and take almost 40 hours of driving time! That's an enormous distance considering a millimeter represents an entire year!

The first three eons of Earth's history are determined mostly by astronomical and geological processes, such as the end of the heavy bombardment of asteroids and comets at the end of the Hadean Eon. However, beginning with the Phanerozoic Eon 542 million years ago, the geological time scales are also determined by life in addition to geological changes to the Earth, a testament to the importance of life on the planet. Very little evidence remains of the first eon, but as we get closer to modern times, we have a more complete picture of the geology, climate, and life's diversity. We can use small clues found in ancient rocks to piece together a picture of the Earth's history. Below is a brief summary of the ages of the Earth, beginning with the first eon and ending with our current epoch.

Hadean Eon (4.6-4 Billion years ago): Named after the Greek god of the Underworld, the Earth in its infancy was a hellish place. Its surface was most likely largely molten and continually bombarded by large meteors. Within less than a hundred million years of its formation, the Earth was hit by another planet-like body nearly the size of Mars vaporizing a large part of the planet that later formed the moon. The surface was almost certainly liquefied and any atmosphere was blown away into space. By the end of the Hadean, the surface had cooled enough to be covered in water and an atmosphere rich in carbon dioxide was present. It's unknown whether or not life was present in the Hadean, mostly because there are no rocks or minerals present from this time period due to tectonic activity and the constant erosion and weathering of the surface.

Archean Eon (4.0-2.5 Billion years ago): Named after ancient Greek for beginning, or origin, the Archean spanned nearly 1.5 billion years of Earth's history. Solid evidence for life exists from colonies of single-celled bacteria called stromatolites, dating back to about 3.7 billion years ago. Other indirect evidence indicates that life may have begun around 3.8-4 billion years ago. Throughout the book, I will use 3.7 billion years as the origin of life, based on the direct evidence. It's important to point out that not everyone agrees with the indirect evidence for the first signs of life. If you were to visit the beginning of the Archean, you would not fare so well because the atmosphere lacked free-oxygen. However, evidence from ancient rocks indicates that cyanobacteria evolved early in the history of life and began to pump oxygen into the atmosphere due to photosynthesis. Over time, these tiny bacteria changed the composition of the atmosphere as free-oxygen became much more abundant. It is generally accepted that volcanism was much more prevalent, bringing lighter rocks to the surface. By the end of the Archean, the first continents existed, but their size and distribution remains largely unknown.

Proterozoic Eon (2500 – 542 Million years ago): Named after ancient Greek for "earlier life", the Proterozoic was the longest eon spanning almost 2 billion years, or 43% of the Earth's history. During this time, atmospheric oxygen levels steadily rose due to the relentless photosynthetic activity of cyanobacteria. Fossil evidence indicates that the first eukaryotic cells and multicellular life evolved during this time, their evolution almost certainly made possible by the rise in free-oxygen levels in the atmosphere. It was during the this eon that the Earth began to take on a recognizable appearance.

Phanerozoic Eon (542-present): Named after ancient Greek for "visible life", the Phanerozoic Eon began with the Cambrian Explosion named after the rapid appearance and diversification of animal life about 542 million years ago. Since the Cambrian Explosion, life has evolved to fill every corner of the Earth. Currently, about 2 million species of life have been described, and the total diversity has been estimated by some to be over 8 million species. Even more amazing, today's diversity represents only about 1% of all the diversity that has existed since the beginning of life. The Phanerozoic Eon is divided into three Eras, the Paleozoic, Mesozoic, and Cenozoic.

Paleozoic Era (542-251 MYA): Meaning "old life", the Paleozoic Era began with the rapid appearance of multicellular life. By 500 million years ago, most modern groups of animals including arthropods, mollusks, and chordates were present. By the end of the Paleozoic, life had greatly diversified and conquered the land for the first time in the Earth's history. Plants had evolved the ability to grow tall, forming vast forest and creating new opportunities for diversification of animals. Fish had evolved and diversified to dominate the seas. Also, one small lineage of fish made the transition to land as they evolved into tetrapods. By the end of the Paleozoic, the early ancestors of reptiles and mammals were present.

The Paleozoic Era lasted for nearly 300 million years, abruptly ending about 251 million years ago with the largest mass extinction of the Phanerozoic Eon. Our current understanding of its end is based on clues left in rocks from the time period. A region known as the Siberian Traps cover an estimated 2 million square kilometers of volcanic rock, which would cover nearly 25% the size of the lower 48 states in rock over a kilometer thick. The massive volcanic activity may have altered the climate and chemistry of the oceans causing the largest mass extinction in the history of life when nearly 85-95% of all species went extinct in a few hundred thousand years, the blink of an eye in geological time.

Mesozoic Era (251-65.5 MYA): Meaning "middle life", the Mesozoic Era followed on the heels of the largest mass extinction the world had ever seen. It took nearly 20 million years for the diversity of life to recover. Despite starting slowly, the Mesozoic Era is best known for its reptiles. The most famous were the dinosaurs that roamed the Earth for nearly 170 million years. The Mesozoic Era also witnessed the evolution of flowering plants, modern mammals, and birds, all of which survived the mass extinction event ending the Mesozoic Era. Evidence suggests that the End-Mesozoic extinction may have been caused by a one-two-punch. It started with a period of massive volcanism altering the climate and the oceans, and ending with the final death blow from a large meteor hitting the Earth 65.5 million years ago, sealing the fate of the conventional dinosaurs along with more than half of the world's diversity. I say conventional dinosaurs because birds are direct descendants of dinosaurs and survived the End-Mesozoic extinction.

Cenozoic Era (65.5 MYA – present): Meaning "new life", the Cenozoic saw the continued diversification of flowering plants, insects, fish, birds, and mammals. Some call it the age of mammals, but there are actually more species of birds than mammals, and there are more species of fish than birds and mammals combined! During the Cenozoic Era, the Earth's climate has slowly cooled resulting in ice ages with large ice caps covering the poles. The Cenozoic Era can be divided into three periods, the Paleogene, Neogene, and Quaternary.

Paleogene Period (65.5 – 23 MYA): The Paleogen Period followed the aftermath of the dinosaur extinction. Diversity recovered quickly at the beginning of the Paleogene. Birds and mammals quickly diversified and competed for terrestrial dominance. Mammals eventually became the top predators and diversified into a great many forms including flying bats, aquatic whales, and terrestrial predators.

Neogene Period (23 – 2.6 MYA): Modern groups of mammals evolved during the Neogene including the ancestors to modern humans. North America became connected to South America when the Isthmus of Panama formed. This connection allowed for a Great American Interchange of life, causing the extinction of many South American species that had been isolated from the rest of the world for tens of millions of years. The isthmus also altered the ocean currents, causing the Earth to further cool and intensifying the current ice age. Not everyone agrees on the start date of the current ice age, some place it starting nearly 34 million years ago when Antarctica became covered in ice sheets.

Quaternary Period (2.6 MYA – Present): The defining events of the Quaternary Period are the growth and retreat of large ice sheets as the Earth oscillated between cold periods of maximum glaciation and relatively quiescent interglacials of reduced ice-cover that we enjoy today. Modern humans also evolved during the Quaternary Period, which is divided into two epochs, and a third that is gaining in popularity.

Pleistocene Epoch (2.588 MYA – 11.7 thousand years ago): Defined by the extensive glaciation of the Quaternary period. It ended with the end of the last major glaciation. Modern humans evolved in Eastern Africa and spread throughout the world.

Holocene Epoch (11.7 thousand years ago – present): Meaning entirely recent, it began at the end of the last major glaciation and encompasses the entirety of written records and human civilization. The Holocene has been a time of relative climate stability.

Anthropocene Epoch (present - ?): Never before has there been a single species on the planet that has had the impact we have unleashed on the Earth in such a very short time. We are rapidly burning fossil fuels, releasing carbon into the atmosphere that had been sequestered for hundreds of millions of years. As a result, we are warming the Earth at unprecedented rates. Combined with the degradation of ecosystems resulting from overexploitation, the spread of invasive species, and rapid climate change, we are causing an extinction event that rivals the end of the Mesozoic era when the dinosaurs were wiped off the Earth. Because of our impact on the Earth, there has been a recent movement among some geologists and biologists to name the current time the Anthropocene Epoch due to the impacts of man's activities on the Earth. At the end of the book, I return to this concept, and suggest that we may actually be entering a new geological era, which I would like to call the Homogenozoic Era.

Throughout the book, I will use these time periods as a reference when I discuss major evolutionary innovations, including the evolution of life, photosynthesis, aerobic respiration, eukaryotes, and the rise and diversification of plants and animals.

Eon	Era	Period	Epoch	m.y.
Phanerozoic	Cenozoic	Quaternary	Holocene	
			Pleistocene	2.6
		Neogene	Pliocene	
			Miocene	23
		Paleogene	Oligocene	
			Eocene	
			Paleocene	66
	Mesozoic	Cretaceous		
		Jurassic		
		Triassic		251
	Paleozoic	Permian		
		Carboniferous		
		Devonian		
		Silurian		
		Ordovician		
		Cambrian		542
	Precambrian	Proterozoic		2500
		Archean		3800
		Hadean		4600

Chapter 2

The Origins of Life

Life began three and a half billion years ago, necessarily about as simple as it could be, because life arose spontaneously from the organic compounds in the primeval oceans.

Stephen J. Gould

Introduction

Perhaps one of the most intriguing questions is what led to the origin of life sometime prior to 3.7 billion years ago. Over the last 60 years, theories explaining the origins of life have continually been proposed, tested, discarded, or modified due to our increasing knowledge of the conditions on the early Earth and biochemical processes. Developing new theories, modifying them, even discarding a popular theory of abiogenesis reflects the process of science by reminding us that science is not static. Instead, science is an iterative and self-correcting process, ultimately leading to a more complete and accurate understanding of the natural world.

In the first half of this chapter, I describe life as a process by asking the question, "What does life do?" It is a subtle difference than describing life as a set of characteristics, but an important difference. By asking what does life do, gets at the heart of this book that life is a process, life does something. A subtle change in how we view life helps us to develop better theories of abiogenesis, understand how life interacts with itself and the environment, and how life has changed over time. I also introduce some basic chemistry, a hard task to sell as something interesting, but I believe a little knowledge of chemistry can go a long way to fully appreciating the process of life. In the second part of this chapter, I cover the history of theories proposed to explain abiogenesis beginning with Darwin. I place them in a historical context to illustrate how scientific theories can change to incorporate new findings.

By now, you may be asking: how do we know our current theories won't be turned over with some new discovery? Is everything we are learning now, just going to be changed in the future? These are valid questions, and sometimes our best theories are overturned almost overnight with a single new observation. But, as science progresses, fewer theories will be overturned and the ones that are correct will become increasingly supported. Science is very good at eliminating hypotheses and theories that don't work.

Life is a Process

One of the most iconic scenes in movie history occurs in Star Wars when a young Luke Skywalker enters the seedy underworld of the Mos Eisley Spaceport Cantina where he encounters numerous alien life forms for the first time. At that moment, his small world expanded as he was introduced to a much larger universe full of strange aliens. Intelligent alien life in this scene may be relegated to the realm of science fiction. But, in real life, we are sending probes to other planets and moons in our solar system and using powerful satellites to search for exoplanets outside the solar system.

To me, finding extraterrestrial life, which is life that did not originate on the Earth, would be one of the most significant events of our lifetime. Finding alien life anywhere in the universe would be humbling, yet a triumphant achievement for humanity to know for certain that we are not alone in the universe. The search for extraterrestrial life begs the question, would we recognize alien life if we saw it? We can answer this question by asking, "What does life do?" If we understand life as a process, then whenever we find these processes, no matter how alien looking, we would be able to recognize it as life.

If we think of life as a process, then at its most fundamental level, life takes in energy from the environment to create order. Life must have energy to exist; without energy available to do work, life could not exist. In fact, if you cut the energy supply off to an organism it will die. Death, like life, is a process where entropy takes over as the complex molecules of cells are broken down into simpler ones and the genetic information that is unique to that living organism is lost forever. Eventually the remains of an organism will reach equilibrium with its environment, and in many cases the elements will enter the geological cycles of the Earth, perhaps to be used again by another animal. Contrary to popular belief, you never want to be in total equilibrium with your environment, because you'd no longer be living.

A necessary feature of living organisms is that they are open systems where energy is constantly flowing through them. For example, energy can enter animals as potential energy stored in the foods we eat. Eventually, most of energy will leave our bodies as heat. However, it requires more than just an input of energy for life to exist. A frozen pizza has every element and molecule required for life. Yet, you can heat a frozen pizza in the microwave all you want and life won't emerge from your efforts to warm dinner. The input of energy is just as likely to break down molecules as it is to build new ones.

Every living organism, from the smallest cell to the largest animal, must interact with its environment to acquire the energy necessary to create order. There is an entire field of science known as ecology dedicated to studying how life interacts with its environment that I will discuss in Chapters 11 and 12. At the base of all ecosystems are organisms called autotrophs (auto = self, troph = feeding) that fix carbon dioxide into organic compounds that are used by all other organisms. The most common autotrophs we encounter are plants that use the energy in sunlight to make sugars from water and carbon dioxide. Autotrophs are vital to life because they transform the energy in sunlight and store it in organic molecules, making energy and nutrients available to other organisms in an ecosystem.

Any organism that obtains their energy from organic molecules, such as the sugars in a piece of fruit, or by eating another living organism is called a heterotroph. Heterotrophs are quite varied and include many bacteria, all animals, and fungi. Recall that energy can't be recycled because every time it is used, it becomes degraded as it is lost as heat, a form of energy that is very difficult to harness to do work. Therefore, energy must constantly flow through living organisms in an ecosystem starting with the autotrophs. Because energy is lost as heat every time it gets used or transferred, there has to be more plants to supply the energy needs to animals. Unlike energy, the elements in an ecosystem, including carbon and nitrogen, are continuously recycled. Decomposers, including many types of fungi, are another type of heterotroph that break down organic matter and recycle it, making it available to plants, who once again use the sun's energy to make the elements available to everything else.

Inside every cell, millions of coordinated chemical reactions take place every second. They break down large molecules into smaller ones and use the smaller building blocks to make larger complex molecules. These reactions also harness the energy released from the break-down of large molecules to do work. You can think of work as using energy to move something. Imagine cleaning your room, by picking up your clothes you are expending energy to do work. However, by creating order inside your room, you actually increase the total disorder of the universe. As you do work, you are using energy, but most of the energy is converted to heat and lost to the environment. That's why you begin to sweat when you are doing work, your body is attempting to remain at the same temperature. The loss of energy as heat whenever we do work satisfies that pesky second law of thermodynamics; every time energy is used for work entropy increases. Although life creates order, it does so at the level of the organism, and can only do so as long as there is a constant supply of energy coming in. So don't worry, life does not violate any laws of nature.

All of those millions of chemical reactions taking place inside of cells are its metabolism, a complex series of chemical reactions used to create order within the cell. These reactions are anything but random, but in fact are highly regulated. A special type of molecule capable of storing information called DNA is used to store the information required to make all the molecules required for our metabolism and to regulate those metabolic activities.

For there to be a continuity of life, there must be reproduction. Without reproduction, life would emerge only to perish shortly afterwards. Reproduction is the process of copying the information stored in DNA and creating a new organism with that information. Sometimes, when information stored in DNA is copied, there are mistakes; these mistakes can be detrimental, neutral, or even slightly beneficial. It is the beneficial mistakes that allow life to evolve over time, continually adapting to a changing environment, or to exploit some new resource. In 1859 Charles Darwin published his book, *On the Origin of Species*, where he presented a lifetime's worth of work supporting the theory of evolution by natural selection explaining how life changes over time to adapt to its environment.

To answer the question, "what does life do?", we can summarize the processes of life. From a physics point of view, life is an island of low entropy, an open system out of equilibrium with its environment. From a chemistry point of view, life is a series of chemical reactions that break down molecules and uses energy to make new, more complex molecules. From a biological point of view, genetic information directing the activities of life is copied to make new organisms, thus ensuring the continuity of life. From a Darwinian point of view, life evolves over time as it adapts to exploit new resources or adapt to a changing environment. At its heart, life is a process that interacts with its environment to extract energy and nutrients to create order.

Lastly, I'd like to return to the question I posed at the beginning of the chapter: would we be able to recognize alien life if we saw it? For me, the answer is yes. If we identify the processes of life elsewhere, then we almost certainly have life regardless of its chemistry or alien appearance. By understanding life as a process, we can also propose testable hypotheses and develop new theories to explain the conditions leading to the first living organisms and look for those conditions on other planets and moons. Ever since Darwin, scientists have speculated and proposed several good theories to explain the emergence of life on Earth. Before diving into the theories for the origin of life, I'd like to discuss the importance of water and carbon for life.

Life Requires Liquid Water

The metabolic processes of life are comprised of countless chemical reactions where molecules are broken down and new molecules are formed. On this planet at least, all of life's chemical reactions are dependent on liquid water to provide a medium for the chemistry of life to occur, simply because water is a good solvent. Without water, the molecules would only rarely encounter each other and life would never arise. One misconception of water is that it is a universal solvent, which is clearly not true. If water were a universal solvent, then it would dissolve anything it touched on contact, including plastic containers.

Another property of water that is extremely important for living organisms is the pH, a measure of how acidic or alkaline the water is based on the concentration of hydrogen ions (H^+). Ions are charged particles; they can be positively charged or negatively charged. Recall that hydrogen has one proton and one electron. If you remove the electron, you have a positively charged ion, in this case specifically, a proton. The pH scale is a measure of the potential of hydrogen ions and ranges from 0 to 14 where 7 is neutral, a pH less than 7 is acidic and a pH greater than 7 is alkaline or basic.

When water is strongly acidic or alkaline, there are more ions in the water, which are very reactive causing chemical reactions to occur. You can see the effects of acidity if you squeeze lemon juice on a piece of uncooked fish. After a few moments, you will notice that the texture of the fish will change as the acidic lemon juice reacts with the proteins forming the muscles in the fish causing them to change shape. The importance here is that chemical reactions and organic molecules have an optimum pH, all cells must exist or regulate pH so they can function.

Water also has several other properties that are important for life. Water is cohesive, it sticks to itself. That's why water is liquid at room temperature, or why water molecules stick together to form a drop of rain. Water also sticks to other objects, a property known as adhesion. Adhesion and cohesion are especially important for plants, it's part of the reason why water can move from their roots to their leaves, sometimes several hundred feet away in very tall trees, without the plant expending any energy to actively move the water.

Water has a high heat capacity, meaning it takes a lot of energy to change the temperature of water, which you may have noticed when trying to boil a pot of water for some hearty macaroni and cheese. The pot will quickly heat up, but the water takes much longer because it is absorbing lots of energy. In fact, water's high heat capacity makes it a great greenhouse gas, capable of holding energy and moderating the air temperature. You may have noticed that cloudy nights don't get as cold as clear nights, that's because water is moderating the temperature. If you've ever lived by the ocean, you may have noticed that the coast doesn't experience the large temperature changes inland areas experience.

The Chemistry of Life is Based on Carbon

Every organic molecule is based on the element carbon. Similar to water, carbon has special features that make it important for life. First, carbon is ideally suited for creating a myriad of complex molecules capable of performing a great many functions including, speeding up chemical reactions, storing information, storing energy, communication, forming barriers, and used in movement, just to name a few examples. The versatility in carbon lies in its ability to form four stable chemical bonds with other elements, including the ability to repeatedly bond to itself, creating an almost endless number of molecules. In fact, the importance of carbon is so great that the entire field of organic chemistry is based on studying carbon compounds.

There may be millions of organic molecules based on carbon, but many of them can be placed into four basic groups of molecule known as macromolecules. They include things you may already be familiar with including carbohydrates, proteins, nucleic acids, and lipids (which include fats). In addition to the macromolecules, there are other types of organic molecules, including vitamins and pigments made by cells that are important for their functioning. Perhaps one of the most important molecules on the planet (there are arguably several others as well) is a green pigment called chlorophyll, which is used to capture the energy in sunlight and convert it potential energy so that it can be used by other organisms. Melanin is a pigment produced by animals and is responsible for eye, hair, and skin color. Vitamins, including vitamin C and B vitamins, are organic molecules that are required for the proper functioning of cells.

Because every living thing is dependent on breaking down and building organic molecules, I have included a brief description of the major functions of each type of macromolecule. Also, an understanding of how these molecules work can improve our health and help us create new medicines to cure cancer or even slow down aging.

Lipids are a broad class of molecules made of mostly carbon and hydrogen and have many important uses. Most commonly, they are known as fats, which are used for long term energy storage, especially by animals. Other lipids known as sterols are used by cells to communicate long distances, two of the most commonly known examples are testosterone and estrogen. Every cell is surrounded by a cellular membrane made of phospholipids, a special type of lipid that has one end that likes water and another end that avoids water, allowing it to spontaneously form cell membranes. Cellular membranes are among the most important structures formed by life. They form a barrier between the living inside of the cell and the non-living outside world. Without cellular membranes, life would remain little more than metabolically active rocks, unable to exists on their own.

Carbohydrates are well known as a source of energy; you may have experienced a surge of energy after drinking a soft drink loaded with sweet tasting sugars including glucose and fructose. Basically every living organism on the planet can easily use glucose as a source of energy, another piece of evidence supporting a common ancestor to all life. By stringing together lots of glucose molecules, complex carbohydrates are made. The most common of these are cellulose and starch, both are made by plants. Starch is used by plants to store energy for later times, the familiar potato is largely starch. Cellulose forms the cell walls of plants and is partially responsible for their structural support. Ironically, cellulose is made of glucose molecules and is the most common organic molecule on the planet, but animals lack the enzymes to necessary to break down cellulose into glucose. Some animals that only eat plants often rely on microbes to breakdown cellulose.

Nucleic acids such as DNA and RNA are used to store our genetic information and regulate the expression of our genes. DNA is a very large molecule made of just four building blocks known as nucleotides (Adenine, Guanine, Cytosine, and Thymine) whose specific sequence codes for every living organism of the planet! Imagine if the English language only had four letters. RNA was once thought to be primarily a message carrier, carrying out the instructions held in DNA to make proteins. However, new evidence suggests that RNA, a close relative of DNA, may have been extremely important in abiogenesis because of its ability to store information and help promote chemical reactions.

Proteins are the work horses of the cell. There are potentially millions of proteins in the world and more are discovered every day. The varied role of proteins is truly amazing, they can be used for structural support, to speed up chemical reactions, cellular communication, and some are used to identify and destroy pathogens. You may be most familiar with the proteins that form your muscles, a common feature of animals that allows us to move around. Although there may be millions of different proteins in the world, they are made of only 20 different building blocks called amino acids. Imagine if the English language only had 20 words!

What Conditions Led to the Origin of Life?

In the 1859 Darwin speculated that life began in a small warm pond millions of years ago. In the 1960s Francis Crick speculated that life may have begun on Mars or somewhere outside our solar system and it was brought to Earth on a meteor, a comet, or even by ancient aliens; a hypothesis known as panspermia. Crick believed that the conditions leading to life were so rare, it was unlikely to have occurred on Earth. However, this scenario is unlikely and the simplest answer is that life began here on Earth sometime about 3.7-4 billion years ago. Part of the reason that it is generally believed that life began on Earth and not elsewhere is because all the ingredients needed for life are right here; a constant supply of energy, liquid water, lots of carbon, and a geologically active planet.

What were the conditions that may have led to abiogenesis sometime around 3.7 billion years ago? Scientists have been studying and testing hypotheses to answer this question for decades. Perhaps one of the most challenging problems regarding abiogenesis lies in trying to reconstruct what the Earth was like at the dawn of life. Was the Earth covered in oceans? Were the oceans salty? Were they acidic or alkaline? Were continents present? What was the composition of the atmosphere? Much of the information has been lost because the Earth is a very dynamic planet with its slow, but steady plate tectonics and the everyday processes of erosion slowly removing the evidence of the Earth's past. As a result, there are often widely different and competing theories describing the conditions of the early Earth.

Perhaps the most widely agreed upon aspect of the Earth's surface 3.7 billion years ago is that it was largely covered in water, much like our current surface. The evidence is based on the fact that the oldest rocks and minerals we have found were clearly formed in a watery environment. Additional evidence from rocks indicates the atmosphere 4 billion years ago most likely was dominated by carbon dioxide, nitrogen, water, and potentially methane, hydrogen, and hydrogen sulfide. The Earth was certainly more volcanic early in its life, which would have added copious amounts of carbon dioxide to the atmosphere. Unlike today, this was good thing because the sun was much younger and only shown at about 75% of its current brightness. The high levels of atmospheric carbon dioxide would have kept the Earth warm, preventing a deep freeze that could have prevented the emergence of life.

All the current evidence from our oldest rocks indicates that the atmosphere lacked free-oxygen (O_2, which is a gas and I will refer to as just oxygen) for the first two billion years or so. In fact, it's unlikely that any planetary atmosphere would have large quantities of oxygen unless something put it there. Although oxygen is necessary for animals, it is extremely reactive, easily breaking down complex organic molecules in to smaller, less complex, and more stable molecules including carbon dioxide. The atmosphere, lacking in oxygen, was conducive to the formation of more complex organic molecules.

The early Earth was most likely geologically active and a warm planet with salty oceans surrounded by a thick atmosphere devoid of oxygen and rich in greenhouse gases. Life emerged through chemical evolution and the direct consequence of geological processes. We also know that the molecules of life are based on the element carbon and the chemistry of life takes place in water. Without carbon or liquid water, it is unlikely that the processes of life could have emerged. Therefore, when looking for places where extraterrestrial life may exist, we would look for liquid water and carbon.

Prior to Darwin, it was thought that life could arise through spontaneous generation, a process where living organisms could simply pop out of non-living material. This was not an absurd hypothesis, but stemmed from observations that fly larvae, commonly known as maggots, seemed to "appear out of nowhere" on rotting plant or animal material. This theory was later experimentally disproved by Louis Pasteur in the early 1860s.

It wasn't until 1953 when Stanely Miller and Harold Urey developed the prebiotic soup theory, providing the first testable hypothesis of abiogenesis. In 1986, the Nobel Laureate Walter Gilbert proposed the RNA world theory predicting that abiogenesis began with self-replicating RNA molecules. In the last decade, a new theory predicting that metabolism evolved before replication, has emerged and is quickly becoming widely accepted because it answers many of the problems with earlier theories of abiogenesis. The continual development and refinement of theories of abiogenesis spanning the last 150 years illustrate how our understanding of difficult problems changes as we gain more scientific knowledge.

Miller-Urey and the Prebiotic Soup Theory

Four billion years ago, carbon was abundant in the atmosphere as carbon dioxide and possibly methane. To determine whether complex organic molecules could form under these primitive conditions, a graduate student named Stanly Miller conducted the famous Miller-Urey experiments with his advisor Harold Urey in the early 1950s. To conduct the experiment, they created a system that contained sea water and a

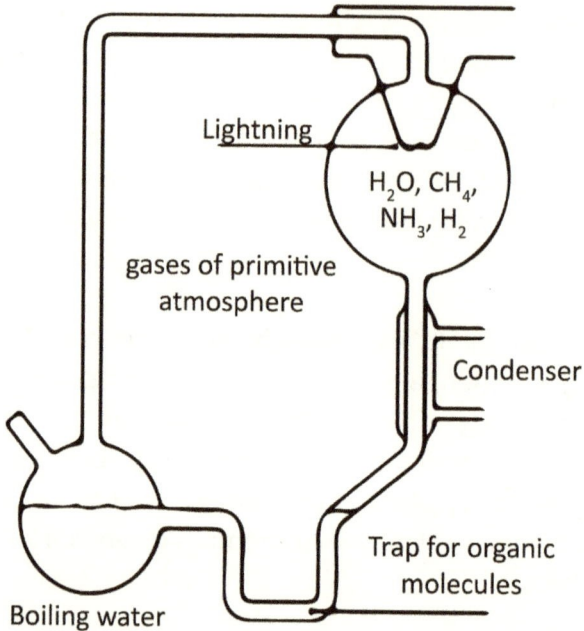

Lightning

H_2O, CH_4, NH_3, H_2

gases of primitive atmosphere

Condenser

Trap for organic molecules

Boiling water

The Miller-Urey Experiment was similar to this set up where water was boiled in flask, the water vapor then entered a gaseous chamber with electrical sparks to simulate lightning strikes, a source of energy. The gaseous mix was then cooled by a condensor and the trap at the bottom collected the organic molecles.

mix of gases they thought represented the early atmosphere. To carry out the experiement, they heated the water and added energy in the form of electricity to the mix. After a week, a brown sludge began to appear in the catchment tube. They collected the substance and found amino acids and other organic molecules that are the basic building blocks of life!

Critics of the Miller-Urey experiments have claimed that they were a failure because they did not produce life. There is a big step going from the building blocks of life to an actual living organism. How could such an improbable event happen? Despite the criticisms of the Miller-Urey experiment, it was a success for several reasons. First, it was the first scientific tests for the origins of life, conclusively demonstrating that the building blocks of life are easily made from simple molecules with an input of energy. The Prebiotic Soup Theory that grew out of these early experiments was a valid scientific theory because it provided a potential explanation for chemical evolution leading to the origins of life. For over 30 years, the prebiotic soup theory was the leading theory to explain the origins of life.

Perhaps the main problem with the prebiotic soup model is that it does not explain how the organic molecules created in the "soup" could come together to form a cell. You can add carbon dioxide, nitrogen, and other small organic molecules to water, add some energy, and you will get the building blocks of life. However, adding energy doesn't mean the molecules will always become increasingly complex, they could just as easily be broken down again. This presents a problem, small organic molecules can be continuously made, only to be broken down again by the next input of energy. Recall that a living organism uses energy to create order, so how would an ordered system, such as a cell, arise from a prebiotic soup of organic molecules?

An attribute of being a scientist is that we continually keep asking questions, forming hypotheses, testing our hypotheses, scrapping or modifying our hypotheses and theories if they are not supported by the data. With new data from observations and experiments, new hypotheses are proposed and tested. From the Miller-Urey experiments of the 1950s, it is known that the building blocks of life are easily formed, an important piece of information for the development of new hypotheses.

The RNA World Theory

The ability to reproduce is one of the key processes of life, ensuring its continuity and existence over time. When life reproduces, genetic information is replicated and used to form the next generation. Based on the importance of replication for life and the findings of the prebiotic soup model, the RNA world hypothesis explaining life's origin became widely popular in the 1980s when the Nobel Laureate Walter Gilbert published a commentary on recent findings in the prestigious scientific journal Nature. Although the theory has its roots dating back to the

1960s, the RNA world is based on the unique properties of RNA, a molecule that is very similar to DNA. RNA, like DNA, is capable of storing genetic information. However, in beginning in the late 1960s and through the 1980s, laboratory experiments showed that RNA is also capable of causing chemical reactions to occur. This means RNA can build new molecules similar to the way enzymes do. What could be better to explain the origins of life? RNA is a molecule that can store information, it can catalyze chemical reactions to make molecules, and catalyze its own replication.

Additional lines of evidence from experiments and observations further support the RNA world theory. One of the important roles RNA has is building proteins in cells. Proteins are literally the work-horses of cells carrying out numerous functions. Because RNA carries genetic information and is directly involved with making the same proteins it carries information for, it has been widely thought that RNA preceded DNA as the first molecule to store information and replicate. The RNA world hypothesis places RNA and replication at the center of life rather than DNA. The reason why DNA is currently used by cells to permanently store genetic information is because it is much more stable than RNA, and it makes sense to store information in a relatively stable molecule.

From its beginning the RNA world hypothesis quickly became a popular theory that has gained wide-spread acceptance. It championed the idea that life began as a simple self-replicating molecule that would eventually evolve into the myriad of life-forms we see today. By using results from the Miller-Urey experiments, scientists knew that the building blocks of life, including amino acids and nucleotides, would form naturally. The RNA world provided an explanation as to how life began starting with self-replicating molecules in a prebiotic soup.

In the last decade, the RNA world has faced several major challenges. Perhaps the biggest question arises from the fact that it is unclear how a large amount of RNA nucleotides could be formed from the prebiotic soup. Or simply stated, what would have been the source of RNA nucleotides, the building blocks of RNA? While RNA is capable of self-replicating, the length a molecule of RNA can reach depends on the concentration of RNA nucleotides. In the prebiotic soup model, the RNA nucleotides would be diluted, and the length of a self-replicating RNA molecule would remain short, much shorter than is required to make even the simplest of proteins. It is also unclear how RNA got inside a membrane and began to direct numerous chemical reactions required for metabolism.

Metabolism First Theory

In true scientific fashion, scientists began to question the RNA world theory as new experimental data began to emerge. Results from laboratory experiments showed that the length of RNA in a prebiotic soup would be too short to form complex proteins. Remember, that based on the principle of uniformitarianism, we know that chemical reactions would operate the same on the ancient Earth as they do today, the laws of the universe don't change over time. That's how we know that experimental results from today can be used to understand the past.

To put forth a new theory of abiogenesis, it has taken new information and the fresh perspective from geologists and biochemists tackling the problem. In the last decade, there have been major gains in our understanding of the chemical properties of the oceans and other geological processes 4 billion years ago. Current geological evidence strongly indicates that the ancient oceans were acidic with a pH less than 7. Today, the oceans are alkaline with a pH averaging 8.1.

A remarkable geological discovery occurred in 1977 when researchers in the deep sea submersible *Alvin* discovered hydrothermal vents in the middle of the oceans. Hydrothermal vents are found near volcanically active areas and spew out hot water along with numerous other chemicals and metals conducive to forming new molecules. Because the water spewing out of these vents is often laden with iron sulfides, it looks like black smoke bellowing from chimneys, hence the name black smokers. Most surprisingly, these hydrothermal vents, found in the cold depths of perpetual darkness of the ocean, were teaming with life. This was a totally new type of ecosystem; instead of using energy from sunlight, they extracted energy from the contents of the hydrothermal vents.

With the discovery of these unique ecosystems surrounding the hydrothermal vents, scientists began to ask whether hydrothermal vents could have been where life began. It is generally accepted that volcanism was much more prevalent billions of years ago making these hot vents commonplace. The hydrothermal vents would have provided energy and a favorable environment for making organic molecules necessary for life to emerge from chemical evolution. The significance of the hydrothermal vents as a source of energy is very important because the first living organisms may have harnessed energy from geological processes to make organic molecules from carbon dioxide and hydrogen gas and other small molecules. It is unlikely that the first organisms on the planet evolved the ability to perform photosynthesis to harness the energy from sunlight.

The discovery of the hydrothermal hot vents was a vitally important clue to life's origins because it provided some answers lacking by the prebiotic soup and RNA world theories. It provided a constant source of energy for early life and a way for organic molecules to be formed and concentrated into small areas. But, like other theories, hydrothermal vents have a couple of problems, the first is that they last only a few decades before becoming inactive. That may not be enough time to allow life to evolve. Secondly, they are very hot, the water coming out them can reach up to 600 degrees Fahrenheit, hot enough to kill almost any living organism.

In 2000, an expedition of scientists, once again in *Alvin*, discovered a different type of hydrothermal vent called an alkaline vent. The most popular one is called the "Lost City", it has about 40 chimneys, some nearly 200 feet high, and perhaps thousands of years old. Alkaline vents are very different in several important ways from the hydrothermal vents previously discovered. Instead of being very hot, they are warm, about 90-110 degrees Fahrenheit. Another major difference is that alkaline vents can remain geologically active for thousands of years, potentially providing enough time for chemical evolution leading to abiogenesis.

THE ORIGINS OF LIFE

In the last decade, William Martin and Michael Russel have suggested that life may have first evolved in similar alkaline vents. Perhaps the most important aspect of the alkaline vents is in their geochemistry. Here's how these vents work, warm alkaline water (pH around 8-9) rich in hydrogen gas and dissolved minerals percolates through the chimneys. Naturally occurring tiny spaces fill the alkaline vents like holes in a sponge forming tiny chemical reactors. The alkaline-rich water would move through these small spaces depositing minerals that are capable of facilitating chemical reactions.

Let's make the connection to the importance of what is happening in these alkaline vents. Recall that a living organism is out of equilibrium with its environment, cells use energy to create order. The alkaline vents would have provided a natural source of energy to facilitate chemical reactions making organic molecules. Because these reactions occurred inside a tiny chamber, about the size of a bacteria, the newly formed organic molecules would have been concentrated there allowing for chemical evolution to take place as organic molecules grew in complexity. This scenario would not violate the second law of thermodynamics because these tiny chambers inside the alkaline vents would have been open systems with a constant supply of energy and materials.

There is more to this story, specifically the importance of the difference in pH between the alkaline vents and the surrounding oceans, which these protocells could harnes as a natural energy source. In the Chapter 3, I will discuss how life uses differences in pH across its membrane to do work.

The metabolism first theory predicts that geological processes caused chemical evolution to occur by creating a system out of equilibrium with the environment. From chemical evolution, life emerged, perhaps starting with metabolism. The implication here is that the first living organisms on the planet may have been little more than a metabolically active rock! Importantly, this theory of metabolism first has been supported by experimental evidence in labs showing that chemical evolution would occur under these circumstances. To date, this is the best theory explaining the origin of life. The ability to use DNA to direct complex chemical reactions and reproduction may have evolved after metabolism, a prediction of the metabolism first theory.

Unanswered questions include how life began to use DNA and RNA as a source of genetic information, a question that scientists are working hard to answer. One of my favorite implications of the metabolic theory is that life originated through chemical evolution, an extension of geological processes. Not only are we stardust, but our distant ancestors may have originated from a metabolically active rock in an ancient ocean!

The First Cells

Life started simple out of necessity. However life got started, it has come a long way evolving into all the plants and animals you see today. Today, there are two basic types of cells based on their structure, the simpler prokaryotic cells and the structurally more complex eukaryotic cells. The first cells on the planet were prokaryotic cells, which were small and mostly lacking internal structures. Today, cell theory tells us that the cell is the basic unit of life and since the origin of life, all cells have come from preexisting cells.

Today, all living organisms share many traits in common, evidence that strongly supports a single origin of life at least 3.7 billion years ago and potentially 100-200 million years earlier. By studying the similarities of living organisms today, we can predict the characteristics of our last universal common ancestor, affectionately known as LUCA. It works because, as life reproduces, it passes genetic information to the next generation. Although, life changes over time, some characteristics have remained little changed since the first living organisms. I'll discuss evolution further in Chapter 4.

One clue to a single ancestor of life is the high energy molecule called ATP, which is used by every cell. By now, you may realize just how important energy is for cells. Actually, cells can't directly use the potential energy in most organic molecules, including carbohydrates, for most the work they need to do. Instead, all cells must transfer energy to ATP so the cell can use it. ATP is used to do practically all the work conducted inside of cells including creating order, or moving the cell around. This molecule is so important that you use about your body weight each day in ATP! In Chapter 3, I will discuss in much further detail how our cells efficiently make ATP. Importantly, it involves chemical reactions and changes in pH, much like the differences in pH between an alkaline vent and an acidic ocean.

Another compelling evidence for a single origin of life is that every living organism stores genetic information in DNA. Even more remarkable is the fact the information to create bacteria, a tree, a fungus, an insect, or even a human is stored in a molecule made of just four different building blocks or the nucleotides, Adenine (A), Guanine (G) Thymine (T), and Cytosine (C). Imagine if the English language only used four letters to compose every song, story, poem, or text book. That's DNA, the story of life is written in the sequence of nucleotides forming DNA.

Even more remarkable is that the language of life, known as the genetic code, is basically the same across all life. It has been inherited unchanged from our common ancestor 3.7 billion years ago. Humans haven't shared an ancestor with bacteria for over 2 billion years. Yet, we can insert any human gene into a bacteria and the bacteria could produce the same protein. In fact, that is how human insulin is made. Scientists isolated the human gene to make insulin, inserted it into a bacteria cell, grew it, and then harvested the insulin. In addition to having a universal genetic code, all living organisms also share similar genes to make the same proteins for similar metabolic processes. One of those involves the break-down of glucose inside cells.

In addition to DNA, there is ribonucleic acid (RNA), a very close relative of DNA as its name would suggest. DNA has one job, it stores genetic information, but it is RNA that carries the information from the DNA to parts of the cell known as a ribosome where proteins are made from their building blocks, the amino acids. Not only does all life make proteins the same way by using RNA to carry information and forming proteins in ribosomes, they also use the same 20 amino acids. The next time you look out the window at a tree, just remember that you are made of the same protein building blocks, share the same genetic code, and actually share similar genes!

Lastly, all life is surrounded by a cellular membrane made of phospholipid bilayers. These bilayers form on their own and create a barrier between the ordered, living component of the cells and the outside world. Life also uses their membranes to make ATP. There are several other characteristics that all living organisms share, one being that metals, like the ones in ancient alkaline vents, are still used to this day to facilitate chemical reactions.

Based on the similarity of all cells today, we can predict that all life had a single origin with certain characteristics. The first cells likely emerged from chemical processes in a warm alkaline vent by using a hydrogen gradient to do work. They created ATP by using a natural gradient across a cellular membrane, used DNA to store genetic information, and used RNA to make proteins from the same twenty amino acids.

Chapter 3

Energy and Life

The "Secret of Life"
May well be found in chemiosmosis
The ability to harness the potential energy
in proton gradients

Introduction

Energy is vital to life. At its most basic level, energy makes things happen, including the processes of life. Energy is a property of matter governed by the laws of thermodynamics, which broadly state that energy cannot be created or destroyed. But, every time you use it, the amount of entropy, a measure of disorder, will always increase. The amount of energy available influences almost every aspect of life, including the evolution of large complex cells and multicellular organisms. Energy allows animals to be active and humans to maintain a very large brain. Therefore, it should be of no surprise that life has evolved efficient ways to extract energy from the environment. Life cannot exists without a continuous flow of energy.

We know life began simply, perhaps it began in an ancient alkaline vent. These unique geological formations would have provided a natural energy source created by the difference in pH between the alkaline water moving through the vent and the acidic oceans. The ability to harness the energy in these proton gradients may be the secret to life. Even today, all cells are capable of generating proton gradients across a membrane to make the majority of their ATP, the universal energy currency of life. It's a process called chemiosmosis that was probably used by the first independently living cells and has been passed down for billions of years to every living cell today.

A common theme in evolution is that structures are often repurposed for other uses, often to exploit new resources. Nowhere is this more true than how cells make ATP by chemiosmosis. It only took small changes to preexisting molecular structures for life to evolve the ability to utilize sunlight as a source of energy for photosynthesis. This small change would greatly alter the course of evolution of life by adding oxygen to the atmosphere. Once again, modifications to preexisting molecular structures would lead to the evolution of aerobic respiration, perhaps the most efficient method of generating ATP available to cells. It was this step that would ultimately pave the way for complex to life, such as plants and animals, to evolve on the Earth.

The Nature of Energy

Energy is similar to life in that it is best defined by what it does; energy can cause things to happen. It is a property of objects that can be transferred or transformed and is subject to the laws of thermodynamics. Energy allows you effect some kind of change in the universe, or more commonly, it allows you to do work. To a physicist, work is the ability to move an object.

There are several ways we can measure energy; the most standard measurement is called the Joule, which is a measure of energy transferred to objects. However, in biological systems, the calorie is commonly used. One calorie is the energy required to raise one gram of water one degree Celsius. Whenever you read calories on food packaging, it is actually in kilocalories or Cal. One kcal of energy could raise a liter of water (about 1 quart) 1°C, or about 1.9°F.

Our understanding of energy was greatly expanded in 1905 when Einstein published his revolutionary Special Theory of Relativity producing perhaps the most famous equation in the world: $E=mc^2$. Einstein's insights regarding the nature energy and mass was revolutionary for the field of physics paving the way for the nuclear age of atomic weapons and nuclear reactors.

The relationship between and mass and energy is explained by Einstein's equation $E=mc^2$ where E = energy, m = mass, and c = the speed of light, a staggering 186,282 miles per second (300,000 km/s). Another way of looking at this is that the mass of an object is also measure of its energy content. To determine the energy content of an object, you would measure its mass and multiply it by the speed of light squared. What this means is that there is an enormous amount of energy in the mass of an object. When the U.S. exploded the first atom bomb in 1945 at the Trinity Site in New Mexico, approximately 1 gram of mass was converted to energy, mostly in the form of heat and light.

Whenever energy is released from a chemical reaction like the burning of natural gas in a stove, a tiny amount of mass is converted into energy. In our everyday world, whenever we use energy, the mass loss to energy is insignificant, measured in millionths to trillionths of a gram, so we won't worry about it here. A consequence of the relationship between mass and energy predicted by Special Theory of Relativity is that you cannot travel faster than the speed of light, it is a universal speed limit. The reason is because it takes increasing amounts of energy to continually accelerate an object. As you approach the speed of light, it would require an infinite amount of energy to push an object past the speed of light.

To use an everyday example to understand why it takes more energy to keep going faster, imagine the difference in energy we spend walking versus running. Most of us can easily walk a mile, to do it in an hour would require little energy. To run a mile in 6 minutes requires you to not only be in good shape, but you also expend much more energy to go the same distance. The faster you run, the more energy you use. It's also why your car gets worse gas mileage at higher speeds, it takes more energy to make your car move faster.

The energy and mass equivalence isn't just limited to physics or science fiction. In biology, animals require energy to move, the faster they run, swim, or fly, the more energy they require. Therefore, there are physiological limits to how fast animals can move and the duration for how long they can maintain top speeds. It's easy to run a mile, but very few people can sprint a mile, and no one can sprint a marathon.

Recall that the laws of thermodynamics govern energy transformations. There are actually four laws of thermodynamics, but in biology, the first two laws are the most important for our purposes. The Earth, all its ecosystems, and living organisms are examples of open systems where energy is constantly flowing through them. In almost every ecosystem, energy enters as sunlight and exits as heat. In contrast are closed systems, such as the entirety of the universe, where energy does not enter or leave.

The first law of thermodynamics tells us that in a closed system, energy is not created nor destroyed. For this reason, it is also called the conservation of energy because the total amount of energy is always conserved in a closed system. The first law also tells us that energy can be transferred from one object to another, and it can be transformed to different types of energy. To illustrate an energy transformation, plants use photosynthesis to transform the kinetic energy in sunlight into the potential energy stored in carbohydrates.

Life being an open system, must maintain a constant flow of energy, or life will decay into equilibrium. We can use ourselves as an example to understand energy transformations in living organisms. All of our energy is obtained in the potential energy stored in the food we eat. Through cellular metabolism that stored energy is eventually transformed into kinetic energy when we go for a walk. The faster we walk the more energy we use. As we break down carbohydrates during exercise, the chemical reactions actually transform most of the potential energy in the foods we eat into heat, another form of kinetic energy. That's why we heat up and begin to sweat when we exercise.

The second law of thermodynamics is nature's reality check for using energy. It states that every time energy is transferred or transformed, or anytime we use energy to do work, some of that energy becomes less useable as the total entropy increases in a closed system. There are a couple of ways to view entropy, the first is a measure of disorder or randomness. If you never clean your house, then entropy would continually increase as it becomes progressively messier. Secondly, as entropy increases, the amount of energy available to do work decreases.

Whenever energy is used, heat is released to the surrounding environment as entropy increases. The more intense you exercise, the the more heat you generate and release to the environment as your energy demands increase. Heat is a very unusable form of energy; it is just the random motion of matter in a system. The more heat, the higher the entropy. Everyday, you are contributing to the total increase in entropy to the universe as you radiate heat from your body.

There are Two Types of Energy

Kinetic energy is the energy of motion, the faster something is moving, the more energy it has. The air around you contains trillions upon trillions of tiny molecules of oxygen (O_2), nitrogen (N_2), carbon dioxide (CO_2) and other trace gases. Each molecule of air has kinetic energy because it is rapidly moving. In fact, the air has so much kinetic energy that it is exerting about 15 pounds of pressure over every square inch of your body. All atoms and molecules have at least some kinetic energy no matter where they are, which is thermal energy or better known as heat. The average thermal energy of a system, such as your body, is measured as temperature. The higher the temperature, the more kinetic energy.

In addition to thermal energy, there are other types of kinetic energy. One we are familiar with is electromagnetic radiation, which includes visible light, microwaves, radio waves, ultraviolet waves (UV), and X-rays. Electromagnetic radiation is made of photons, tiny particles that travel in waves at the speed of light. When you step into the sunlight, you can feel the energy as heat on your skin. Stay in the sun long enough and ultraviolet light will begin to damage your skin cells, accelerating the aging process. Other types of kinetic energy can be found in sound waves, which are are made when something vibrates. Unlike light waves, sound waves cannot travel through a vacuum, including outer space (you would never hear a spaceship exploding in space!) Electricity is the movement of electrons from one place to another, representing another form of kinetic energy.

Potential energy is any type of stored energy. As an animal, we get all the energy we need as chemical energy stored in the bonds of molecules. It takes kinetic energy to break chemical bonds and energy is released when new chemical bonds are formed, the process is a chemical reaction resulting in new molecules. Watching a campfire is a great example of energy transformations that you can actually watch. Potential energy is stored in the cellulose of the wood. If you add a little kinetic energy in the form of a match, a chemical reaction begins when the chemical bonds holding carbohydrate molecules together are broken and new chemical bonds are formed making carbon dioxide and water. When the new bonds are formed, potential energy is transformed to kinetic energy releasing photons of energy that you see as light and feel as heat.

When there is a lot of potential energy stored in a molecule, the chemical bonds are less stable and are easily broken. Some molecules like octane, a component of gasoline, contains lots of potential energy and the chemical bonds are easily broken. A small spark supplies enough energy to break the chemical bonds, igniting the gasoline. Once this happens, the chemical bonds of the gasoline are rapidly broken as they quickly form carbon dioxide and water. The potential energy stored in the gasoline is quickly converted to kinetic energy in the form electromagnetic radiation that we see as light, thermal energy that we feel as heat, and sound waves that we hear as the boom.

Molecules, such as water and carbon dioxide store very little energy because they are stable, the chemical bonds holding them together are difficult to break. If you've ever thrown a match into water, it simply goes out. The flame lacks sufficient kinetic energy required to break apart the water molecule into its elements, oxygen (O) and hydrogen (H).

Similar to kinetic energy, potential energy comes in different forms. We are familiar with gravitational energy, the higher you place an object, the more potential energy it contains. If you were to drop a glass only a few centimeters from the floor, little would likely happen. Drop the same glass from a few meters in height and it will likely shatter. And of course, water stored in large lakes behind a dam is also a form of potential energy.

We learned from Einstein that an immense amount of energy is tied up in in the mass of objects. Inside our sun, nuclear fusion slams hydrogen atoms together forming helium, and in the process, releases vast amounts of energy. However, in the last 4.6 billion years, the sun has only lost about 0.008% of its mass, or about the equivalent of Earth's mass. Nuclear energy is stored in the nucleus of atoms. Some isotopes are radioactive and decay into smaller elements releasing kinetic energy. In nuclear reactors, the heat generated by radioactive decay is harnessed to generate electricity.

Cells spend a lot of energy moving ions across their membranes. As they do so, one side becomes more positively charged than the other side forming a membrane potential. This is very similar to how a battery stores energy. Because there is both an electrical gradient and a chemical gradient due to the different concentrations of ions across the cellular membrane, they are commonly called electrochemical gradients.

The energy stored as an electrochemical gradient across cellular membranes is incredibly important, allowing them regulate how much water is in the cell or the movement of other materials in and out of the cell. And, as I've stated earlier, cells can also use electrochemical gradients to make ATP. Based on the importance of membrane potentials to cells, it has been hypothesized that the first cells may have gotten their start by harnessing the potential energy of naturally forming electrochemical gradients found in thermal vents.

Chemical Reactions
Transfer and Transform Energy

Life interacts with its environment to acquire the energy necessary to create and maintain order. Organisms can be broadly classified as autotrophs or heterotrophs based on how they acquire energy from the environment. Autotrophs are capable of using energy from their environment in the form of light (photosynthesis), or potential energy from inorganic chemicals, minerals, and compounds (chemosynthesis) to make organic molecules. Basically, autotrophs add hydrogen to carbon dioxide to make complex organic molecules. Heterotrophs must acquire energy from the environment as chemical energy stored in organic molecules made by autotrophs. As an animal, we are heterotrophs that acquire potential energy from the foods we eat.

Inside our cells, energy-rich organic molecules, including carbohydrates and fats, are broken down and their energy is transferred to do work, creating order within cells. We are most familiar with work as the ability to move an object. In addition to powering our muscles, life at the cellular level spends a lot energy moving objects across their membranes to maintain an internal environment that is different from outside the cell, known as homeostasis. Every day, your body spends about 20% of its energy pumping ions (mostly sodium, potassium, and calcium) across membranes to create electrochemical gradients as part of the process of maintaining homeostasis. Cells also use energy to make larger more complex molecules from smaller building blocks. All of these chemical reactions inside a cell, breaking molecules down and building larger ones, is our metabolism, which consists of both exergonic and endergonic reactions.

Exergonic Reactions: The energy cells use for movement, maintaining homeostasis, or building larger molecules, comes from exergonic reactions that release energy. Although the organic molecules we consume in our food are rich in stored energy, most are relatively stable at room temperature. They simply do not randomly burst into flames breaking down into smaller molecules. To break the chemical bonds between the elements in a molecule, an input of energy is required called the activation energy. The more potential energy stored in a molecule, the less activation energy is required.

To illustrate this example, wood and paper are made of cellulose, a complex carbohydrate made of numerous glucose molecules, each one comprised of carbon, hydrogen, and oxygen. If you've ever started a fire with a match, you are supplying the activation energy required to break the chemical bonds between the elements in the cellulose. Once the original bonds are broken, the elements want to quickly form new bonds so they will be chemically stable. Below is the general chemical equation for cellular respiration, which is also similar to the burning of wood.

$$C_6H_{12}O_6 + 6O_2 \longrightarrow 6H_2O + 6CO_2$$

You can see that there are different molecules on each side of reaction. One molecule of glucose ($C_6H_{12}O_6$) and six molecules of oxygen (O_2) on the left side represent the reactants. On the right side of the equation are the products, six molecules of carbon dioxide (CO_2) and six molecules of water (H_2O). It is also a balanced chemical equation because the same number of elements are on each side; for example, there are six carbon atoms in the reactants and in the products. Chemical reactions do not change the elements.

Recall that electrons are important in forming chemical bonds and oxygen really likes electrons. When organic molecules are broken down in the presence of oxygen, we can say that they are oxidized, a type of chemical reaction that occurs when oxygen takes away electrons. Oxidized molecules are often smaller and store less potential energy than reduced organic molecules. Cellular respiration, or the burning of wood has increased entropy by not only by releasing heat, but also because the products are simpler and less ordered than the reactants. Inside of your cells, similar reactions are continuously take place, transferring energy and releasing heat to the environment.

Without an input of energy, exergonic reactions, while chemically favorable, may occur very slowly, or not at all. Because oxygen really likes electrons, every once in a while, it will react with sugar or other organic molecules to form water and carbon dioxide. But, the molecules are shy, they have to contact the glucose molecule in exactly the right way, or nothing happens. As a result, sugar sitting on your table lasts a long time. It would eventually break down, but it may take decades to do so.

Endergonic Reactions: Endergonic require an input of energy and create order inside cells by building larger and more complex molecules. Cells use some of the energy from exergonic reactions to supply the energy necessary for endergonic reactions. There are several reasons why an endergonic reaction requires an input of energy. First, the simpler molecules of their reactants may have strong bonds that are hard to break. So more energy is used to break the bonds of the reactants than energy is released when the products are formed. Second, entropy is being lowered inside the cell by the creation of larger more complex molecules from more simple ones. As a result, the products on an endergonic reaction contain more potential energy than the reactants.

Recall that exergonic reactions supply the energy required to build complex molecules. However, the energy released from the exergonic reaction must always be greater than the energy required for the endergonic reaction, satisfying that pesky second law of thermodynamics. Without a constant supply of energy coming into the cell, endergonic reactions would cease, and exergonic reactions would eventually cause the break-down of all the large organic molecules as the cell decays towards equilibrium with its environment. The process happens even more quickly in the presence of oxygen. Below is the general equation for photosynthesis, a type of endergonic reaction.

$$6H_2O + 6CO_2 \longrightarrow C_6H_{12}O_6 + 6O_2$$

Plants are autotrophs that fix carbon dioxide into carbohydrates, an energy intensive process requiring a constant supply of energy to break the strong chemical bonds in water and carbon dioxide. The reactants in photosynthesis are oxidized, it is difficult to split them apart because the oxygen is holding on the electrons. That's why plants rely on the energy in sunlight for this reaction. When carbon is fixed into an organic molecule, we say that it is reduced, and it has gained electrons and potential energy. Also notice that there are fewer products and more reactants.

Making ATP: The Energy Currency of Life

All life uses the same energy currency, a high energy molecule called adenosine triphosphate, better known as ATP. The more ATP available inside a cell, the more work that can be done. Energy entering a cell is typically not directly used by the cell, instead it is transferred to ATP. This goes for both plants and animals and every other living organism. ATP is so important to cells, that some cellular biologist have called it the molecule of life. Every day, you use about your body's weight of ATP to carry out all of your body's functions from breathing, running, studying and storing memories, or producing every protein in your cells. The universality of ATP among all life is strong support for a universal common ancestor to all life on Earth from bacteria, to plants, to humans.

A molecule of phosphate is comprised mostly of electron grabbing oxygen atoms that all attain a slightly negative charge. With three phosphate groups attached to the adenosine molecule, the negative charges of the oxygen repel each other, storing lots of potential energy, and making the molecule somewhat unstable. When ATP is used, the last phosphate group comes off forming adenosine diphosphate or ADP. The removal of the third phosphate group is an exergonic process, releasing energy and the resulting ADP molecule has less potential energy.

The regeneration of ATP from the less energetic ADP is of paramount importance for cells and they have evolved two basic ways to accomplish this task, substrate level phosphorylation and chemiosmosis. Substrate level phosphorylation uses a protein enzyme to add a phosphate group to ADP, forming ATP. Because the process requires energy, it is directly connected to the energy released from the break-down of organic molecules. It is an example of energy coupling where cells transfer the energy released from one reaction to another.

Perhaps the most commonly used source of energy for substrate level phosphorylation comes from a 10-step metabolic process called glycolysis, which literally means 'splitting sugar'. It is the first stage of a much larger series of chemical reactions known as cellular respiration, which I will discuss in more detail later in this chapter. Because nearly every cell is capable producing some ATP through glycolysis, it is thought to have evolved early in the history of life and is further support of a single-common ancestor for all life.

Despite the fact that most organisms are capable of forming ATP by glycolysis, it is not the most efficient method. For example, it takes an exact amount of energy to make ATP, about 7.3kcal/mol. If an exergonic reaction releases 12.3 kcal/mol, then the extra energy is lost to the environment as heat. Or, if you don't have enough energy, you don't make any ATP, and the energy simply lost to the environment. A 'mol' is an abbreviation for a mole where one mole of an element would weigh its atomic weight in grams. Because atoms are tiny, it takes a lot of them to equal their weight in grams, in fact about 6.023×10^{23}, or a little over

600 billion trillion atoms of carbon-12 to weigh 12 grams.

Another way to better understand the reason why combining two reactions together to make ATP is not very efficient is to imagine purchasing a soft-drink that costs $1.50/20 ounces. If you put in $5 a vending machine, you got one soda, but you lost your $3.50 in change. On the other hand, if you put in $1.49 in change, no soda for you, and you lose the money you put into the vending machine! The same principle applies for making ATP. If a reaction releases 7.2kcal/mole of energy, you do not get any ATP, that energy is simply lost to the environment.

Chemiosmosis: the Secret of Life

If you've ever wondered if there is a "secret to life", it could be chemiosmosis; the ability to harness proton gradients to efficiently make ATP. Chemiosmosis is a word which literally means "to push". It is nature at its most efficient, harnessing the potential energy stored in electrochemical gradients across a cell's membrane. A simple way to think about gradients is that they are a change in some value across a distance. If you've ever walked up a hill, then you have gone up an elevational gradient, where the elevation changes over the distance you walked. An electrochemical gradient has two parts; the electrical gradient formed by different charges across the membrane, much the same way that a battery stores potential energy as a voltage, the difference between the "+" and "-" end. The chemical gradient is formed by the difference in ion concentrations across a membrane.

Recall that a proton is the same thing as a positively charged hydrogen ion (H^+). When there are more protons on one side of a membrane a chemical gradient is created. An electrical gradient is formed because there are more positively charged protons on one side of the membrame. Together, an electrochemical gradient is formed storing potential energy. Because there is a gradient, the system is out of equilibrium. The protons want to be in equal concentrations on each side of the membrane. But, the ions can't easily cross a cellular membrane, which restricts their

Chemisosmosis

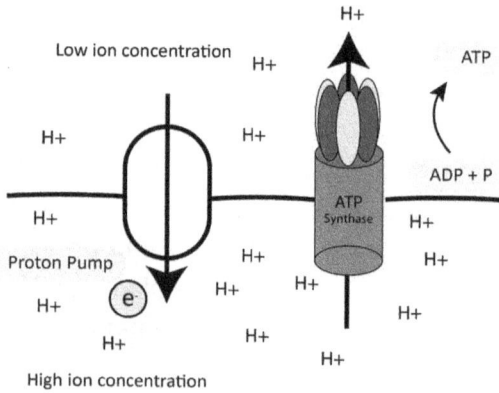

Chemiosmosis is an efficient way of making ATP by using the potential energy found in proton gradients.

movements. It is similar to water behind a dam, the water can't move through the concrete, so it builds up to higher levels behind the dam, storing energy.

To harness potential energy stored in the electrochemical gradient, the ancestors to all cells evolved a large protein complex called ATP synthase. The 'ase' ending means this is an enzyme that helps a chemical reaction take place. When lots of protons build up on one side of a membrane, they "push" their way through the ATP synthase to the other side of the membrane in an effort to reach equilibrium. As the protons push their way down their electrochemical gradient, the ATP synthase uses the kinetic energy of the moving protons to make ATP.

Chemiosmosis is efficient at making ATP because energy can be stored in a gradient across a membrane until there's enough energy to make a molecule of ATP. It's like putting your change in a piggy bank, once you have $1.50, you can buy your soda from the vending machine and you don't waste your money.

Based on our current theories of life's origins and early evolution, the next major evolutionary leap for life occurred when cells were able to use other energy sources to generate their own proton gradients. This step was crucial because it allowed cells to live independently of their place of origins and spread to new areas and exploit new resources, such as the energy in sunlight.

The answer for these cells was the evolution of large protein pumps powered by electrons that actively moved protons across cellular membranes, generating their proton (electrochemical) gradients. The first organisms capable of generating their own proton gradients likely obtained their energy and electrons from various sources in their environment. In the earliest autotrophs, or self-feeding organisms, the source of electrons likely came from minerals or rocks. In heterotrophs, they obtained their electrons from organic molecules. While there are many sources of electrons and energy to power the proton pumps, there must also have been electron acceptors capable of removing electrons, or the whole process would have stopped as it reached equilibrium.

Today the series of proton pumps and electron carriers is collectively called the electron transport chain. Together with ATP synthase the entire process effeciently makes ATP through chemiosmosis, a process that is ubiquitous among all living cells. Overtime, the electron transport chain has evolved to use different sources of electrons, energy, and electron acceptors.

If ATP is the molecule of life, then chemiosmosis is one of the most crucial metabolic processes of life. Based on the ubiquitous nature of chemiosmosis, it's generally accepted that for more than 3.7 billion years, some form of chemiosmosis has been preserved in every organism, inherited from the first living organisms on the planet. In the next two sections, I explain how cells have repurposed the molecular machinery of chemiosmosis to use new sources of energy and electrons in photosynthesis and a new electron acceptor in aerobic respiration.

Photosynthesis
Utilizing an Abundant Energy Source

The evolution of photosynthesis was one of the most important evolutionary innovations in the history of life. Photosynthesis is a word that literally means 'putting together' (synthesis) with 'light' (photo). It's a process that takes the energy in sunlight and stores it in organic molecules, mainly carbohydrates. Small prokaryotes called cyanobacteria were most likely the first group of organisms to evolve photosynthesis perhaps some 3.5 billion years ago. Organisms capable of photosynthesis are autotrophs, which are capable of using energy from their environment to form organic molecules from carbon dioxide.

Photosynthesis is a remarkable evolutionary innovation for two reasons. First, these organisms had access to a practically unlimited source of energy from the sun, they no longer relied on energy from natural proton gradients or from acquiring organic molecules in their environment. Second, these ancient cyanobacteria could harness the energy from the sun to create all the organic molecules they would ever need. That's why photosynthetic organisms are at the base of almost every ecosystem in the world; they make carbon and energy readily available to the rest of us. To illustrate the importance of photosynthesis, just think that every carbon atom in your body was once part of a carbon dioxide molecule in the atmosphere that was fixed into an organic molecule through photosynthesis. Therefore, photosynthetic organisms form the basis of almost every ecosystem, making energy and nutrients available to all other life.

The chemical equation for photosynthesis may appear relatively simple, but in reality it is a multistep process requiring sunlight, water, carbon dioxide, many proteins, and cellular structures dedicated to the process. There are two main parts of photosynthesis, the light reaction and the synthesis reaction, also known as the Calvin cycle. Typically, we all learn about photosynthesis as it takes place in the chloroplasts inside plant cells. Chloroplast are tiny organelles, or tiny cellular organs that were once free-living bacteria. Plants never evolved the ability to perform photosynthesis, instead, the ancestor of all plants actually acquired the ability to conduct photosynthesis by engulfing cyanobacteria. Over time, they evolved into the chloroplasts that are found in plant cells.

Visible Light: Before we dive into photosynthesis, it's good to know a little about the properties of light. Visible light is actually a part of the electromagnetic spectrum where tiny packets of energy called photons, travel in waves. The shorter the wavelength, the more energy in the photons. We see a tiny fraction of the electromagnetic spectrum as visible light where the wavelengths of red light are about 700 nanometers and blue light is about 380 nanometers. Ultraviolet light has a wavelength ranging from 100-380 nanometers and carries enough energy to damage your skin. Infrared light that we can feel as heat has a wavelength from about 800 nanometers to about 1 millimeter. A nanometer is one billionth of a meter. To put that tiny size into perspective; if you were to enlarge an object that was one nanometer to a millimeter in length, (about the width of the letters you are reading) it would be the same as making a meter stick one thousand kilometers (about 600 miles) long!

The Light Reaction: Whether or not photosynthesis is taking place in a cyanobacteria or in the chloroplasts of a plant cell, the basics of photosynthesis are similar. It begins with the light reaction, which accomplishes several tasks. Importantly, it tranforms the energy in sunlight into a usable form of chemical energy in ATP throough chemiosmosis

How does the energy in light waves get transformed and temporarily stored in ATP during photosynthesis? In the light reaction, the proton gradient required for chemiosmosis is created by using electrons to power proton pumps as they travel down an electron transport chain. Does this sound familiar? The transformation of light energy to ATP in the light reactions is very similar to how the first cells made ATP by using proton gradients and chemiosmosis. Remarkably, it was probably not a giant evolutionary leap for life to evolve the light reaction of photosynthesis. A recurrent theme in evolution is that structures are continually repurposed for new tasks. In the case of the light reaction, electron transport chains were repurposed to help transform light energy to ATP. Although the source of electrons and energy is different for the light reaction.

Inside photosynthetic cells, the light reaction takes place in large molecular complexes called photosystems embedded in cellular membranes dedicated to the light reaction (you can think of a photosystem as a large satellite dish to collect light). Its purpose is twofold; first to use the energy in sunlight to split water apart, ripping the electrons away from oxygen. The second purpose is to elevate the electrons in energy so they can be used in an electron transport chain to generate a proton gradient. A second photosystem re-energizes the electrons after they have lost their energy traveling through the electron transport chain.

The light reaction begins when chlorophyll, the pigment that makes plants green, absorbs and concentrates light energy. Once the light energy has been "captured" in the chlorophyll, some of that energy is used to split water into hydrogen and oxygen. The oxygen exits the cell as free oxygen (O_2), a very important consequence of photosynthesis that I will return to later in this chapter. Water is tough to split apart because oxygen likes to hold onto its electrons. Therefore, plants need the energy in sunlight to split water by pulling electrons away from oxygen. Also, the electrons that came from the splitting of water lack energy and at this point, can't be used to power the proton pumps.

The second purpose of the photosystem is to use light energy to excite the electrons in energy. Once the electrons have been elevetated in energy, they go through a series of electron carriers and proton pumps collectively called the electron transport chain that creates an electrochemical gradient by pumping protons across a membrane, ultimately storing energy.

Imbedded in the same membrane that houses the photosystems and the electron transport chains are ATP synthases. To reach equilibrium, the protons "push" their way through the ATP synthase and in the process, ATP is generated through chemiosmosis. Recall that this is the same process of chemiosmosis used to generate ATP in practically every cell on the planet. The energy in sunlight has now been transferred to potential energy stored in ATP, which will be used as an energy source for the Calvin cycle. After the electrons have traveled down the electron transport chain, they have once again lost their energy. At this point, a second photosystem energizes the electrons so they can be used to fix carbon dioxide into organic molecules in the Calvin cycle.

To recap the light reactoin, it begins when sunlight strikes the chlorophyll pigments in a photosystem, the energy is concentrated as

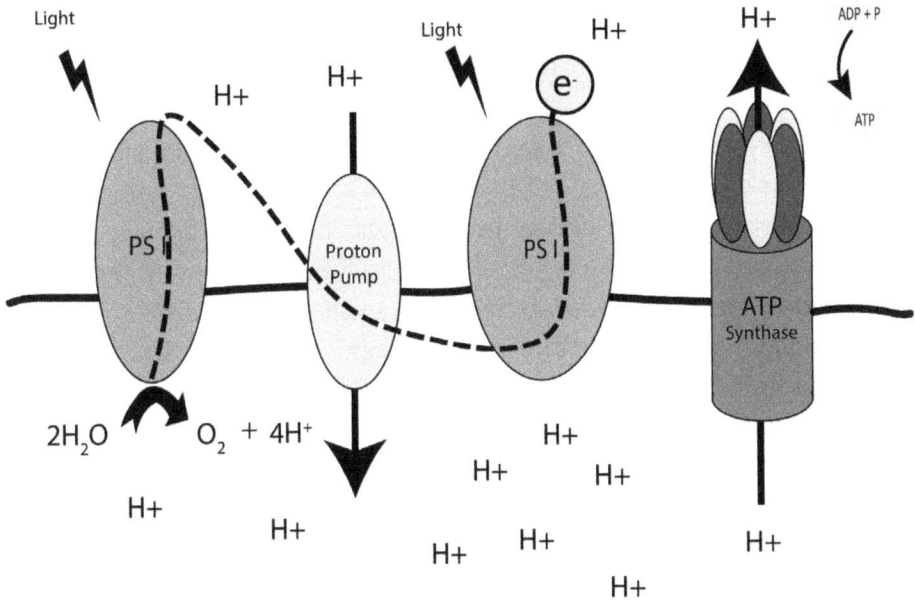

Photosynthesis uses the energy in sunlught to remove electrons from water and elevate them in energy so they can power proton pumps used to create a proton gradient. The potential energy is used by ATP synthase to make ATP through chemiosmosis.

water is split and electrons are energized. These high energy electrons are used to power proton pumps, generating the proton gradient used to make ATP through chemiosmosis. A second photosystem, re-energizes the electrons so they can be used to make organic molecules in the Calvin cycle.

The Calvin Cycle: The Calvin cycle is a series of chemical reactions that takes carbon dioxide (CO_2) and fixes it to organic molecules by adding hydrogen atoms to it. Carbon dioxide is very stable, or inert, it doesn't really react with other molecules. It takes a lot of energy to make an organic molecule out of carbon dioxide because its difficult to break the chemical bonds between carbon and oxygen. The energy for the Calvin cycle comes directly from energized electrons from the light reaction and the ATP produced by chemiosmosis. The source of hydrogens and electrons to be added to carbon dioxide comes from the water that was split in the light reaction. Once carbon has been fixed, it bceoms available to life. In fact, every carbon atom in your body was fixed in the Calvin cycle.

Aerobic Respiration
An Evolutionary Response to Free-Oxygen

The free-oxygen released from photosynthesis energized life on the planet. Sometime between 1.5 and 2 billion years ago atmospheric oxygen levels reached a critical point, it was becoming toxic to the life forms that had existed in a relatively oxygen-free environment for almost two billion years. A few researchers have even gone so far as to call the rise in oxygen at this time the Great Oxygen Catastrophe because they speculate that the rapidly rising oxygen concentrations may have caused a mass extinction of microbial life not adapted to oxygen. Even today, there are bacteria, known as anaerobes, which will die in the presence of oxygen as they have no defenses against its oxidizing tendencies. In reality, it probably was not so much of a catastrophe for life as it was an evolutionary opportunity.

Life has an uncanny ability to evolve and adapt to its surroundings, allowing it exploit new resources, including oxygen. To use the energy locked up in organic molecules, heterotrophs must transfer it to ATP. There are several ways cells can do this, and one that they all use is chemiosmosis, similar to what takes place in photosynthesis and in the first cells. In a remarkable twist of fate, the same electron grabbing property that makes it damaging to cells can be used greatly improve the efficiency of chemiosmosis, making it possible to make much more ATP from the same amount of organic molecules.

When oxygen became more abundant, some prokaryotic cells evolved small modifications to their electron transport chains giving them the ability to use oxygen as the final electron acceptor, removing them from the chain. This one small evolutionary step was a major leap forward for life. First, it added the protons that were used in chemiosmosis to the oxygen molecules and formed water; in one simple step the toxic effects of oxygen were neutralized. But, the real leap was in the fact that oxygen has a higher affinity for electrons than all the other electron acceptors previously used by life. This meant that the electrons could be used to do more work because of the bigger energy drop they could undergo. This larger energy drop generated bigger proton gradients allowing cells to extract much more energy from organic molecules, thus producing much more ATP from every organic molecule they consumed. Cells now had more energy to grow larger and become more active.

Based on molecular evidence, the first aerobically respiring bacteria may have been the proteobacteria, a large group of bacteria that include pathogenic species such *Salmonella* and *Helicobacter*. They also include the familiar *E. coli* found in our guts and some bacteria that fix nitrogen. Eventually, an aerobically respiring bacteria merged with another prokaryote and togehter they would go on to evolve into eukaryotes, a totally new type of cell. Inside almost all eukaryotes are the descendents of the proteobacteria, which are the mitochondria. they evolved into energy-converting organelle responsible for aerobic respiration and the vast majority of ATP production in all plants and animals. The evolution of eukaryotes is known as endosymbiosis and I will discuss in Chapter 7.

In our cells and all other eukaryotes, aerobic respiration has three major steps; glycolysis, Krebs cycle, and oxidative phosphorylation, which includes the electron transport chain and chemiosmosis. The first step, glycolysis is the only step that does not take place inside the mitochondria. It's important to note that glycolysis and the Krebs cycle are the center of cellular metabolism.

Glycolysis is basically splitting (lysis) sugar (glycol) in half. It is a catabolic pathway requiring 10 steps, with each step using a different protein enzyme to make each reaction happen. Glycolysis is not limited to eukaryotes. It occurs in practically every cell on the planet and may be one of the oldest metabolic pathways because it does not require the presence of oxygen or specialized structures inside the cell. It only produces a net of 2 ATPs for every molecule of glucose, not very much ATP considering the amount of potential energy in sugars. A n o t h e r fascinating aspect of glycolysis is that its end product, called pyruvate, can serve as a precursor for many other molecules in the cell including amino acids, lipids, and nucleotides; all the building blocks of life.

The Krebs cycle was named after Hans Krebs who spent ten years working out the steps of this cycle in the 1930s. Similar to glycolysis, it also occurs in many types of cells, but is mostly studied inside of mitochondria, which are only found in eukaryotic cells. In cellular respiration, the overall purpose of the Krebs cycle is to break down organic molecules by stripping them of their hydrogens and electrons. This cycle only produces 2 ATP, similar to glycolysis. However, the electrons carry energy and are used for the final step of aerobic respiration. The carbon atoms that entered the Krebs cycle are released as carbon dioxide that simply diffuse out of the cell and exhale them as you breathe.

Oxidative phosphorylation is the final step of aerobic respiration and it occurs on a membrane similar to photosynthesis. The high energy electrons stripped form organic molecules travel down an electron transport chain and are used to power large proton pumps protons to generate an electrochemical gradient. The protons then flow through ATP synthase to produce ATP by chemiosmosis. This is essentially the same process we encountered in the light reaction of photosynthesis.

Inside the electron transport chain, the electrons flow from one protein complex to the next, doing a work at each step. By the end of the electron transport chain, the electrons have lost their energy and are hard to remove. Because oxygen really likes electrons, it is able to pull them off the electron transport chain, ensuring a continuous flow of electrons, ultimately, maintaining a proton gradient for chemiosmosis. Because the electrons all end up on oxygen at the end of aerobic respiration, we say that oxygen is the final electron acceptor. Not only does oxygen pick up the electrons, it goes on to react with the protons coming through the ATP synthase forming water.

Does the process of oxidative phosphorylation sound familiar? Here is the third example of ATP being produced by chemiosmosis, a process utilizing a proton gradient generated by an electron transport chain and protons flowing through ATP synthase. In fact, the ATP synthase found in plants and cyanobacteria is very similar to the ATP synthase found in the mitochondria of all eukaryotes. As you can see, chemiosmosis has

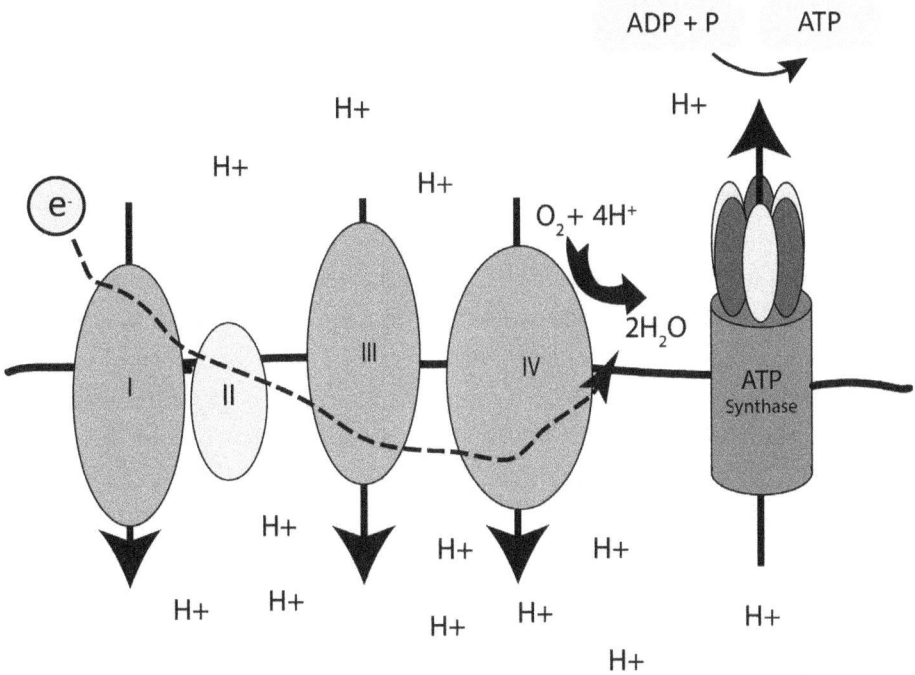

Oxidative phosphorylation is similar to photosynthesis, they both use electron transport chains to generate a proton gradient that powers the production of ATP as protons flow through an ATP synthase. Organic molecules serve as both a source of energy and electrons for the process.

been repurposed several times over the course of evolutionary history. In photosynthesis it was repurposed to transofrm energy from sunlight in ATP. In aerobic respiration, it was repurposed to protect cells from the oxidizing effects of oxygen with the added benefit of producing more ATP.

Organisms that undergo anaerobic respiration don't use oxygen and produce far less ATP per organic molecule. This inefficiency ultimately places limits on the size and complexity of organisms. Without the presence of oxygen, larger more complex eukaryotic cells would never have evolved. Life would have remained as simple unicellular organisms.

Life Changed the Earth Forever

Photosynthesis changed the world in many ways. To begin, every carbon atom in your body was once in a molecule of CO_2 in the atmosphere and was fixed into an organic molecule by the Calvin cycle. In fact, adding hydrogen atoms to carbon atoms to form organic molecules is fundamental to the existence of life; every organic molecule contains carbon and some hydrogen. To date, there are only about 4 metabolic pathways, including photosynthesis, that are capable of performing such an important chemical feat. Three of those are limited moslty to unique geological features or are not capable of growing very fast. If photosynthesis had never evolved, there probably wouldn't be the abundance of life on the Earth we see today.

Photosynthesis had another major effect on life. The byproduct of photosynthesis is free oxygen (O_2), a very reactive molecule. Oxygen as an element was abundant on the early Earth, but it was bound up in other molecules such as carbon dioxide. There was no free-oxygen in the atmosphere because it would quickly react with other molecules whether they were in the water, rocks, or newly formed minerals.

Over geological time, free-oxygen was continually produced by photosynthesis and eventually oxidized almost all the iron in the oceans forming banded iron formations. Once the iron was removed from the oceans, they became the blue color we are accustomed to today. It's hard to imagine, but the world looks much the way it does today due to oxygen in the atmosphere, a byproduct of photosynthesis! Additionally, we mine those ancient banded iron formations to make steel.

Oxygen had a huge impact on the atmosphere, the oceans, and every rock that has formed on the surface in the last 2 billion years. In fact, if scientists were to find an exoplanet with more than 1% oxygen, it would be a smoking gun for photosynthesis, which is a biological process and clear indication of extraterrestrial life. As the oxygen levels increased in the atmosphere, it also had a profound impact on life.

Animals did not evolve aerobic respiration, bacteria did, just like they evolved photosynthesis and practically every other major metabolic pathway used by life. Sometime, about 2 to 1.5 billion years ago, the first eukaryotic cells evolved in response to the higher levels of oxygen. The ensuing evolution of eukaryotic cells was the largest restructuring of cells in the history of the planet. This new type of cell was larger, had numerous mitochondria efficiently producing ATP, a cell nucleus that housed the DNA, and the insides became additionally compartmentalized into organelles. In fact, the name eukaryote means "new kernel" because these cells contained a nucleus unlike the smaller prokaryotes.

Another major consequence of atmospheric oxygen is that high up in the atmosphere, it reacts with ultraviolet light to form ozone (O_3). Ozone is even more reactive than oxygen and is a pollutant at the surface. But, high up in the atmosphere it blocks incoming ultraviolet light, protecting the surface from these damaging high-energy light waves. Without the thin layer of ozone, life could not exist on the surface, it would be killed by the ultraviolet light. Life would be relegated to watery environments where it would be protected from harsh ultraviolet light.

About a billion years ago, a eukaryote engulfed a cyanobacteria, which evolved into a chloroplast, the organelle in modern plants responsible for photosynthesis. This new cell contained both mitochondria and chloroplasts meaning they were autotrophs that could perform photosynthesis, but they also performed cellular respiration efficiently generating ATP.

Eventually, some eukaryotes began forming colonies that evolved into the first multicellular organisms about 600 million years ago. One group went on to form plants, another into fungus, and of course, another into animals. Just think, life had been around for 3 billion years before the arrival of animal life! It would take another 150 million years for animals to begin colonizing the land. It appears that plants, fungus, and animals all began colonizing the land as early as 500-450 million years ago. By this time the Earth was already 4 billion years old.

Today, all multicellular organisms including plants, fungus, and animals are comprised of eukaryotes descended from those two prokaryotic cells that came together evolving into one cell. Prokaryotic cells have never evolved into multicellular organisms and perhaps never will. By now, I hope that you see the connections between energy and life. When bacteria repurposed an electron transport chain for chemiosmosis, evolving photosynthesis to utilize a source of constant energy, they produced oxygen, a by-product that was detrimental to many organisms. Once again, life evolved and adapted by once again repurposing the electron transport chain for chemiosmosis. This time, it reacted oxygen with hydrogen producing water, but in the process, the production of ATP was vastly improved. Life was energized, and with more energy available to life, we see the evolution of multicellular organisms, higher diversity, and more complex ecosystems.

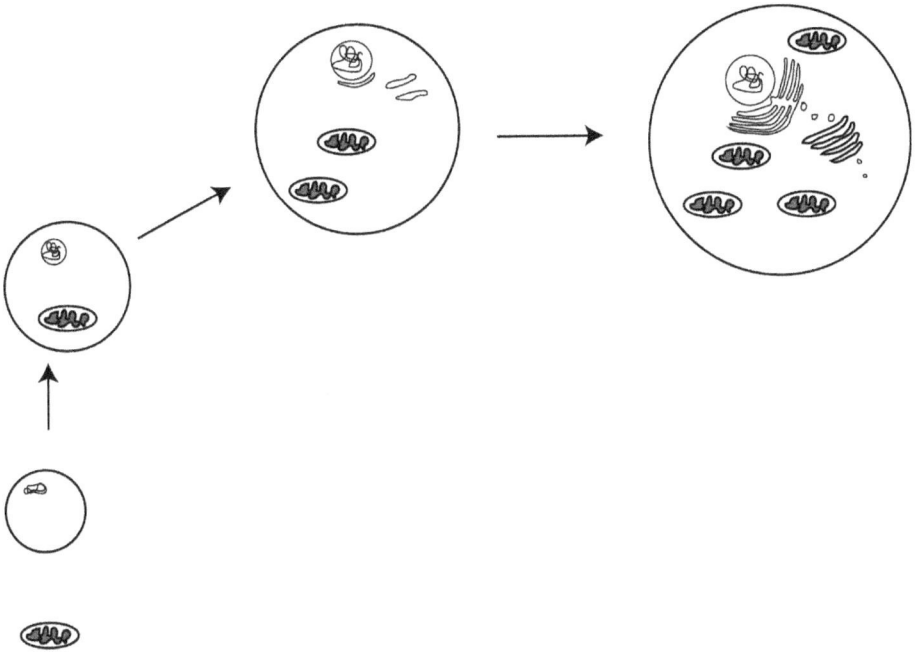

Eukaryotic cells evolved through a process called endosymbiosis. About 2 billion years two prokaryotes merged to form one cell. The smaller prokaryote was an aerobically respiring bacteria, capable of using oxygen to effeciently make more ATP. The larger prokaryote may have been a methane producing Archaea. Eventually the bacteria would evolve into mitochondria and the overall size of the cell would grow as various organelles evolved compartmentalize the functions of eukaryotic cells. About a billion years ago, a eukaryotic cell engulfed a cyanobacteria, acquiring the ability to conduct photosynthesis.

Chapter 4

Scientific Revolutions Evolution by Natural Selection

To raise new questions, new possibilities, to regard old problems from a new angle, requires a creative imagination and marks real advance in science.
Albert Einstein

To deny scientific discoveries is to deny reality. Nothing in biology makes sense except in the light of evolution
Theodosius Dobzhansky

Introduction

Scientific discoveries of the past few hundred years have revolutionized our understanding of how the natural world works. There have been many debates over which discoveries were the most important, were the most ground-breaking, or which scientist was the smartest, or made the biggest contribution to the advancement of human understanding. I'm not going to weigh in on that debate too much because major advancements were all made by incredibly knowledgeable scientists who dedicated their lives to learning. They were all intellectually gifted with imagination and creativity allowing them to elucidate the workings of the natural world.

At its heart, this chapter is about evolution by natural selection, a scientific advancement that like Newton's *Principia* or Einstein's theories of relativity, revolutionized the way we understand the natural world and conduct science. The impact of Darwin's theory reaches well beyond biology to include all fields of science. Upon its publication in 1859, it fundamentally changed the way we do science, forever removing any need to invoke supernatural explanations to understand the natural world. By the 1940s, evolution was firmly established as the central paradigm to biology.

Surprisingly, biology is the only major field of science that a large portion of the general public does not believe in its central paradigm. In fact, many revolutionary theories are met with resistance from other scientists, philosophers, and the general public because they challenge our deeply held beliefs, anecdotal knowledge, and often wrong common sense knowledge about how we think the world works based on our limited observations and experience. Always keep in mind that science is a process about learning how the natural world works, it's based on evidence from experiments and observations. Science moves forward as we gain knowledge; hypotheses are proposed, tested, discarded, modified, or accepted. It's an iterative process, based on the works of many people, and always improving our understanding of the natural world.

Scientific Revolutions

Revolutionary theories can take decades to be accepted by other scientists and even longer for society. That's because some theories are just too revolutionary based on the current knowledge and the *zeitgeist* of the time. *Zeitgeist* is a German word that means the 'spirit of the times', or the intellectual fashion of the time. Beginning with the Renaissance, philosophers and gentlemen naturalist began relying more on observations and experiments to advance science, rather than just accepting the conventional wisdom of their times. However, their early scientific discoveries often contradicted common sense knowledge of the time, making it difficult for people to accept their new findings because it directly challenged the accepted world view.

More than once, scientific discoveries have gone against the prevailing thoughts of their time. For most of the existence of humanity, it was thought that the Earth was flat and the universe revolves around the Earth, giving credence to an inflated importance of man's place in the universe. This predominate world view was based on simple observations that the sun moves across the sky and the Earth doesn't seem to move if you are standing on it. Additionally, the surface of the Earth appears flat from where you are standing.

The first challenge to the flat Earth hypothesis dates back over 2250 years ago to the Greek Philosopher Eratosthenes who predicted that the Earth was round. He used a simple observation that two poles in different locations, one in Alexandria and the other in Syene, cast shadows of different lengths at the same time of day. Using basic geometry, Eratosthenes correctly calculated the circumference of the Earth to within a few hundred miles, which was a very accurate measurement.

Eratosthenes work was largely ignored for 17 centuries because it contradicted common sense knowledge when everyone thought the world was flat. He was finally proven correct in 1521 when the crew of Ferdinand Magellan successfully sailed around the Earth, proving once and for all it was round. But, people still thought the Earth was the center of the Universe. However, in 1543 a major challenge to this view was put forth when Copernicus published his theory of heliocentrism, predicting that the Earth and planets revolve around the sun. Once again, few people accepted his work because it was such a radical departure from the prevailing view of society placing the Earth at the center of the universe. Copernican heliocentrism eventually became widely accepted with additional discoveries a century later from Galileo.

Using the first telescope, Galileo observed moons orbiting Jupiter, lending additional support to the notion that everything in the universe does not revolve around the Earth. The importance of Copernicus's and Galileo's heliocentric theory was that they used observations to develop testable models of the solar system, ultimately advancing our world view. Their heliocentric theory was extremely controversial because it removed the Earth as the center of the universe, reducing the anthropocentric view of man's place and importance in the universe. Today, we have satellites that routinely take images of the Earth against the vastness of the universe proving the Earth is round and it orbits the sun. The long-held belief of society based on "common sense knowledge" that the world is flat or the is the center of the universe has been totally disproven by the relentless progress of science. Today we now know that the Earth is but a tiny blue speck in the vastness of the cosmos.

Another great scientific revolution occurred in the 1680s when Newton published his laws of motion and gravitation in one of the most important scientific books in European history, *Principia*. Newton's laws of gravity and motion were the first forces of nature to be clearly articulated through a mathematical treatise. In these laws, he described the effect of gravity on an object by stating that any object dropped on the Earth would fall at the same rate of 9.8 m/s^2. The same laws of gravity and motion were also used to describe the orbits of the planets beyond the Earth. By doing so, Newton clearly demonstrated that the fundamental forces of nature are the same everywhere. This thinking was also in stark contrast to the prevailing thoughts of the day in Europe that stemmed from Aristotle believing that the laws of science were different beyond the moon, or the celestial realm.

Newton ushered in a new era of physics by making the characterizing of fundamental forces a priority of physics. Newtonian physics of gravitation and motion became the dominant paradigm of physics for the next 200 years. Although, Newton didn't explain gravity itself; that would take another two centuries and a profound intellectual leap in imagination and creativity by another great mind. Newton's laws were so successful that by the 1890s, a few physicists were starting to believe that most of the laws of physics were about to be discovered. That naive world view was shattered when a relatively unknown physicist named Albert Einstein developed his revolutionary theories of relativity.

In 1905 Einstein published his Special Theory of Relativity describing the mass-energy equivalence in his famous equation $E=mc^2$. This theory ushered in the atomic age and forever changed our perception of space and time. Perhaps, the most difficult aspect of special relativity is that it challenges our "common-sense" view of the universe by stating that there is no standard reference point to the universe because the speed of light is "fixed". Here's another way to explain this, imagine you are standing on the road and a car passes by at 50 mph, if someone from the car throws a ball at 20 miles an hour, then the ball would be traveling at 70 mph relative to the person standing on the roadside. Einstein's remarkable insight was that this common sense approach does not work for light; rather, he showed that light will always travel at the same speed, 186,282 mph, relative to the observer. Meaning, if I turn on a flashlight, the speed of light will be 186,282 mph. If a car passes me at 100 mph and turns on the headlights, the light will be traveling at 186,282 mph for the car and for someone standing on the roadside! The speed of light is constant regardless of the relative motion of the source!

In 1915, Einstein once again devised another revolutionary theory, the General Theory of relativity, which explains gravity. The origins of General Relativity date back to Newton's Law of Gravity, which uses math to characterize the effects of gravity on objects including the planets, but did not explain how gravity worked. Cracks in Newton's laws of gravity were present in the early 1900s. Using Newton's laws of gravity, the orbit of the planets were pretty well worked out; except for Mercury, the closest planet to the sun. It turns out that its orbit was not exactly as it should be based solely on Newtonian physics. This was an indication that Newton's Law of Gravity may not be the whole story to gravity.

Einstein's General Theory of Relativity predicted that massive objects, like the sun or the Earth, actually curve space-time, creating a "valley" in the fabric of the universe that objects would fall into. Taking into account the curvature of space-time by the sun, Einstein's General Theory of relativity correctly predicted Mercury's orbit. Another verification of his theory came in 1919 when astronomers were able to verify that light from a distant star was bent as it went near the sun. This independent verification of Einstein's theory made him an instant household name. However, it would take several more decades for the full impact of General Relativity to be appreciated because it also predicted black holes, the Big Bang, and an expanding universe before there was a single observation supporting any of these predictions. Over the past 100 years, Generally Relativity has been repeatedly verified by numerous observations and experiments. It has been proclaimed by some notable physicists that Einstein's General Theory of Relativity may have been the biggest leap of scientific imagination in history. Surprisingly, Einstein never received a Nobel Prize for either of his theories of relativity.

There are other great scientific discoveries that have had a profound influence on science, some more controversial than others. In the 1800s, the laws of thermodynamics that govern energy transformations were developed to build steam engines. In the 1920s quantum mechanics was developed to explain the bizarre universe at the tiny scale of atoms where an electron can appear to be in two places at once, or act as a wave and a particle, or where subatomic particles can instantaneously influence each other over great distances, a process known as quantum entanglement. Although this may sound esoteric and of little practical use, our knowledge of quantum mechanics makes lasers and transistors used for modern computers possible. The Big Bang theory that grew out of Einstein's equations and Hubble's observations of expanding galaxies was thought to be so absurd by one famous astronomer that he called it the "Big Bang" to ridicule it.

Each one of these discoveries was born out of observations, experiments, imagination, and creativity. They have been repeatedly confirmed through extensive experiments and observations. It takes a lot of evidence and support to create a good scientific theory. In fact, the iterative and self-corrective nature of science either modifies or discards theories if they do not hold up to the evidence. And lastly, the imagination and creativity of these scientists were vital in providing new insights into the natural world or devising unique methods to test theories. Einstein, perhaps one of the greatest minds to have ever existed in the world, once proclaimed that imagination is more important than knowledge. Indeed Einstein was special case considering he devised his revolutionary theories as thought experiments while working as a patent clerk!

A lesson from Einstein's Theories of Relativity, Copernicus's heliocentrism, and especially the strange behavior of tiny subatomic particles explained by quantum mechanics is that the natural world often defies our "common sense" views. How we commonly perceive the day to day world is not necessarily how it actually works. These discoveries should be a reminder that we must use the results of our observations and experiments to understand the natural world and not rely on what we think is true.

Darwin's theory of evolution by natural selection was another revolutionary theory that has been met with skepticism and outright denial. For many people the fact that species can change over time defies common sense knowledge based on a very limited time frame. The problem stems from the fact that our lives are much too short to witness the large evolutionary changes in species taking place over millions of years. The events of our lives unfold in days, years, and even decades, and the entirety of human civilization is only a few thousands of years old. So short are our lives that we are ill-equipped to fully understand geological timescales of millions of years, let alone a 100 million years, and a billion years is basically unfathomable. In that amount of time, a lot of evolutionary changes can take place.

To put into perspective the length of our lives in comparison to geological time, our fish-like ancestor slowly evolved to live on the land about 375 million years ago. From there, it has evolved into mammals, reptiles, birds, and amphibians. To understand the vast amount of time it has taken for modern humans to evolve, it's helpful to use our road trip example of using distance to understand time. If you were to go on a road trip back in time to witness the first vertebrates on land, using 1 mm = 1 year, you would have to drive 375 km, or about 233 miles to observe our ancient ancestor adapting to live on land. However, the span of our entire life may be less than three inches.

However, just like the theories of Eratosthenes, Copernicus, and Einstein, Darwin's theory of evolution by natural selection has been well supported by more than 150 years of repeated verification from thousands of scientists. Since its publication, evolution by natural selection has become the central paradigm of modern biology and is considered one of the greatest intellectual contributions to modern science. Darwin's theory changed the way we do science by removing the need to use supernatural explanations to explain the natural word.

The Time Was Right for Evolution

Sometimes a scientific theory is proposed at the right time, by the right person, and they receive the majority of the credit for the discovery. Darwin is credited for developing the theory of evolution by natural selection, although he was not the only person to develop this theory. However, he was the right person, in the right place, at the right time to put forth a revolutionary theory that remains with us rather than becoming a side-note in history. Eratosthenes provides a great example of a person who was ahead of their time, most people have never heard of this Greek philosopher, even though he was the first person to demonstrate the Earth is round. It's just that his work was 1500 years too early!

Sometimes, it can take a long time to change long-held views of society put in place by earlier people. Prior to Darwin, the prevailing western scientific thought for centuries was that the world was young and all the species were created perfectly in the recent past. Many naturalists before Darwin's time, including Linnaeus, erroneously believed that each species of plant or animal was created perfectly in the recent past, and any deviation would be a deviation from perfection. A consequence of this dogmatic view of the natural world was that it slowed scientific advancement for centuries.

Unfortunately, this dogmatic view of an unchanging world slowed and even prevented scientific progress, even though it was not supported by observations and experiments. Aristotle had a huge influence on western thought for almost two millennia. For example, it took nearly 1800 centuries for Galileo to conduct a simple experiment to disprove Aristotle's hypothesis that heavier objects fall faster than lighter objects, once again an example of science disproving common sense knowledge of the world. Another wrongly-held belief that lasted about as long was that women had fewer teeth than men. Once again, it only took one person to count the number of teeth between men and women to dispel this idea.

If you've ever watched a nature program only to hear the narrator state that some unique animal is "perfectly adapted", then you've witnessed Aristotle's thinking about perfection in nature. The reality is that there are no perfectly adapted organisms because there is always a trade-off. If you can run fast, then you lack power; if you are perfectly camouflaged, then you may not be able to move fast.

Beginning in the early 1800s, cracks to the unchanging world view were beginning to emerge from geology. Two geologists, Hutton and Lyell, were proposing theories of uniformitarianism to explain the formation of modern geological features of the landscape such as mountains or valleys. Uniformitarianism stated that the same geological processes operating today, operate in the same in the past. For example, if a mountain is growing one inch per year, then most mountain ranges will grow at approximately the same rate. Or if erosion slowly carries away sediment from the land, it will do so at similar rates in the past. The clear implication of uniformitarianism is that it predicts geological processes are slow compared to the life of a human; it takes millions of years to build a mountain or to erode one down. These two early geologists were among the first to believe the world was ancient, perhaps tens to hundreds of millions of years old.

An ancient Earth contradicted many beliefs in the 1800s, including the young Earth idea predicting the age of the Earth to be a few thousands of years old. At the time, this was not an absurd idea because no one really had any good way of dating the Earth so there was no real reason to believe the Earth was ancient. It wasn't until geologists proposed the principle of uniformitarianism to explain some of their observations, including the presence of sea shells on mountains thousands of feet above sea level. They realized that geological forces were very slow, but over time, it could account for fossils of marine organisms on high mountains. By Darwin's time, the age of the Earth was predicted to be approximately 10-100 million years old. In the late 1890s radioactivity was discovered giving scientists a tool capable of determining just how ancient the Earth really is. Although, it would take another 60 years to fully understand radioactive decay and half-lives of elements, but eventually the Earth was determined to be approximately 4.6 billion years old. Far older than anyone had predicted or thought possible.

Coinciding with the advent of uniformitarianism in geology, the first scientific hypotheses to explain the origin of species, or how species change over time, were being proposed. They were born out of numerous observations by naturalists and museum collectors who were combing the world, searching for strange new animals and fossils to be catalogued for museum collections. In the early 1800s, the French naturalist, Lamarck published the first scientifically coherent theories of evolutionary theory by stating that the environment causes changes in animals. However, his mechanism of change, acquired characteristics, was later disproved by observations and experiments. Despite proposing an incorrect mechanism for evolution, Lamarck was the first person to use biology as we do today. Ironically, Erasmus Darwin, Charles Darwin's grandfather, also proposed the idea that species could change over time.

Adding to the progress made by naturalist and geologists of time, the economist Robert Thomas Malthus, published a series of articles on population growth between 1798 and 1826. He showed how a population could grow exponentially if there were no deaths from disease, famine, or other causes. He believed that unchecked population growth could lead to catastrophe. Malthus's work would later become influential in Darwin's formulation of evolution by natural selection where he clearly made the point that in nature, more individuals are born than can survive.

By the 1850s, the young Earth idea was being routinely challenged by geologists as it became clear that geological processes are slow. Naturalists were collecting and studying numerous fossils, plants, and animals from around the world, bringing them back to be described and catalogued. Coinciding with advances in natural history, the works of Malthus and Lamarck were being debated in scientific circles. It appeared that the time was right for the right person to put all these pieces together into one revolutionary theory.

Darwin's Journey to Discovery

Charles Darwin was born on 12 February 1809 into a well-to-do family in England, the same day Abraham Lincoln was born. What a great day for the world! Darwin originally started his education by attending medical school, but he quickly lost interest after watching surgery on a young child without anesthesia. Rather, he showed an early propensity to collect plants and animals, similar to the naturalist of the time. In the summer of 1831, one of Darwin's professors presented him with an opportunity to be a gentlemen's naturalist and proper companion to Captain Fitzroy of the HMS Beagle. The purpose of the journey was to map the South American coast over a two-year period. At first, Darwin's father was not supportive, calling it a waste of time. Luckily, Darwin's uncle convinced the elder Darwin to let his son go and even to financially support the endeavor.

The HMS Beagle left port for South America on 27 December 1831. While on the journey, Darwin was given a copy of Lyell's *Principles of*

Geology, which ended up having a big influence on Darwin's views on the age of the Earth. In fact, on the Beagle's first stop at the Cape Verde Islands, Darwin noticed sea shells in geological formations several hundred feet above the surrounding Atlantic Ocean. This was one of the first signs of the Earth's ancient age to the young Darwin, providing solid evidence to back Lyell's theories. While at sea, Darwin often suffered bouts of sea sickness. Fortunately, the Atlantic Ocean offered calmer periods when he spent time collecting plankton, dissecting marine invertebrates, and preparing specimens to be shipped back to museums. And importantly, he spent time writing copious amounts of notes detailing his observations and findings, which would later prove invaluable in his later writings of evolution.

Once the Beagle reached the shores of South America, Darwin spent most of his time on land collecting plants, animals, fossils, and minerals. While Darwin made his collections, the Beagle carried out its primary mission, surveying the South American coastline. At times, Darwin would spend several months exploring South America before being picked up by the Beagle to sail further down the coast. During one of his month-long treks near Patagonia in present day Argentina, Darwin discovered the fossils of large extinct mammals; providing another piece of evidence indicating that species arise, change over time, and go extinct.

Perhaps some of the most important clues indicating a changing and ancient Earth occurred when Darwin was exploring in modern day Chile. Along the western edge of South America, the Andes Mountains rise thousands of feet, forced up by the collision of the South American plate into the Pacific Plate. Hiking up into the Andes, Darwin saw fossils of sea shells thousands of feet above sea level, far higher than the sea shells he witnessed on the Cape Verde Islands. While in port, he noticed that a mussel bed had been recently lifted above sea level after a recent earthquake, much like the fossilized mussel bed he observed in the Cape Verde Islands at the start of the voyage. His logical conclusion was that given enough earthquakes over a long period of time, mountains could rise carrying sea shells once at the bottom of the ocean to heights thousands of feet above sea level. Later this observation would also lead him to believe that the Earth was old enough for species to slowly change over vast amounts of time.

Perhaps the most famous stopping point of the Beagle's journey were the Galapagos Islands located about 600 miles off the coast of Ecuador. These are geologically young islands, less than 5 million years old, with a unique assemblage of plants and animals on each of the main islands. Darwin noticed that although the plants and animals were unique species to the Galapagos, they closely resembled similar species in South America. For example, there were flightless cormorants, giant tortoises, and marine iguanas all resembling similar species to the mainland 600 miles away. As the Beagle sailed across the Pacific Ocean, they stopped at other islands allowing Darwin to make additional collections and observations of other unique island plants and animals. Over time, a pattern began to emerge, all the oceanic islands contained unique species; however, island species most closely resembled species to the nearest mainland. Taken together with the evidence of an ancient Earth, these observations inspired Darwin's view that perhaps species could change over time.

The Beagle continued on to Cape Town in South Africa and then back across the Atlantic to finish a few surveys of the South American coast before returning home to England. What was to be a two-year mapping expedition of South America, turned into a five-year journey that circumnavigated the world. Finally, the Beagle returned to the same port in October 1836. Upon arrival, Darwin had already gained fame as a naturalist based on his collections he had sent back during his five-year journey. He would spend the next few years writing books about his adventures and the natural history of his collections. In the ensuing decades, he would use his collections, additional observations of selective breeding in cattle, pigeons and crops, along with the influence of other geologists, naturalists, and the works of Malthus to develop his theory of evolution by natural selection. He was goal was to accumulate enough information and form a well-supported theory that explained his observations, including those of the unique island species that were similar to ones on the mainland.

Darwin's Theory of Evolution by Natural Selection

By the 1850s, it was the right time and Darwin was the right person to bring forth the theory of evolution by natural selection explaining that all modern species descended from common ancestors. Although, Darwin may have been intentionally delaying publication or even considered the idea that he would publish his theories posthumously because he knew that it was going to incur rancorous debates among the scientific elite and would be ridiculed by the public. Even Darwin's close friend, Charles Lyell, had encouraged him to publish the work or risk getting "scooped".

This almost happened when in 1858, a young naturalist named Alfred Russel Wallace sent Darwin a manuscript detailing essentially the same theory of evolution by natural selection that Darwin had been working on for decades. Darwin had Wallace's work presented alongside his at the Linnean Society that year. In 1859, Darwin published his book, *On the Origin of Species*, which quickly sold out. Darwin is generally given credit for evolution by natural selection because he had clearly been working on the theory for decades and presented copious amounts of support for it. Additionally, there was a clear paper trail indicating that he had developed the theory prior to Wallace.

While many naturalists immediately accepted Darwin's work, it still took decades to become fully accepted, partly because no one understood inheritance. Eventually, advances in our understanding of genetics solidified evolutionary theory. In the end, it revolutionized the way we do science and became the central paradigm of the entire field of biology. Darwin's theory that species changing over time, slowly evolving with modifications from ancestors, simultaneously explained the unity and diversity of life that we observe today. The basic tenants of the theory have been routinely verified by repeatable observations and experimental data for over 150 years.

Species change over time, this is a fact that is verifiable by observations and experiments. We can see it happen and measure the rate species change either in the controlled environment of scientific laboratories or in less predictable natural settings. But the question remains, why do species change over time? If species do change over time, what kind of predictions can we make to test this theory? The theory of evolution by natural selection, like any good theory, explains the observations made by scientists and generates additional testable predictions.

Life is abundant on the Earth, you can find it on the land, in the sea, or buried deep in the ground. The variety of life in our world is truly stunning, there are whales measuring 100 feet long, trees that live for thousands of years, lichens capable of eking out a living from minerals on rocks, and simple prokaryotes living in water hot enough to boil. Everywhere you look, life is present and is incredibly diverse. Yet, with all the diversity, there is an underlying unity revealing evolutionary relationships as modern species descended from common ancestors.

Darwin, among others, including the great taxonomist, Carolus Linnaeus, observed that there were similarities among species. In fact, Linnaeus developed the modern classification of life based on shared characteristics between different taxonomic groups. For example, all animals with hair and mammary glands are classified as mammals based on those shared characteristics. If these mammals possessed a flap of skin between their limbs and were capable of powered flight, then they would be classified as bats.

Science begins by making observations, sometimes they are new observations that no one else has noticed. Darwin, like other naturalists had made several key observations about populations. Similar to other great scientist that put forth revolutionary theories, Darwin did not develop his theory of evolution by natural selection in a vacuum, he was influenced by his friends and other scientists, and from reading the works of other scholars. First, he observed that there was variation among individuals within a population. You can easily see this just by looking around a classroom or walking into a crowded area, no two people are the same. Second, he observed that more individuals are born in a generation than can survive, which coincides with the work of Malthus. Malthus was an economist who developed the principle that unchecked populations would grow continually at an exponential rate. For example, if a pair of fruit flies reproduced and all their offspring survived, and reproduced, and this continued for each generation, then the world would be covered 3 feet deep in fruit flies in less than four months! His work was highly influential on Darwin, because clearly, we are not drowning in a vast sea of fruit flies.

One attribute that makes a scientist great is their imagination and creativity that allows them to make new insights or see things in a different way from their contemporaries. Darwin observed that there is variation among individuals within a population and that more individuals are born than can survive. Darwin's insight was that survival and reproduction are not random. Evolution by natural selection predicts that the individual that is best fit to its environment is most likely to survive; but most importantly, it is more likely to reproduce, passing on those favorable traits to the next generation. Over time, species change to become better adapted to their environment as they accumulate favorable traits for that environment. Hence the common phrase, survival of the fittest. Darwin used his creative insight to put together his observations into a single testable theory explaining numerous observations regarding the unity and diversity of life. It's important to note that natural selection acts upon the individual, but only populations are capable of adapting and evolving. An individual cannot evolve.

Natural selection is a theory because it provides a mechanism explaining how species change over time. A logical conclusion of this theory is descent with modification; species change over time from ancestral species. Therefore, modern species share traits inherited from a common ancestor. For example, all mammals have hair and mammary glands similar to the first mammals that evolved those traits. Likewise, if you were to find a new animal that had hair and mammary glands, then you could classify it as a mammal. Hence, the theory explained the unity of life through descent from a common ancestor.

Today, there are about 5,416 known species of mammals. They are a diverse group including whales, bats, cats, and humans. The diversity lies in the fact that the first mammal appeared about 240 million years ago in the Triassic Period, and mammals have been diversifying into different habitats before the dinosaurs went extinct, a very long time! Once again, evolution explained the diversity of life as they descended with modification from a common ancestor.

The Fossil Record Provides Great Support

Evolution by natural selection is fully supported by the fossil record as it provides a clear record of descent with modification. While in South America, Darwin discovered fossils of large extinct mammals. Combined with geological support of an ancient Earth, it became apparent that species would come into existence and then go extinct or evolve into new species. He predicted that in time, transitional fossils, or fossils that have characteristics of ancient species and modern species would be found. His prediction was correct when the fossil of Archaeopteryx was discovered in Germany around 1861. Archaeopteryx was a transitional species with traits of both dinosaurs and birds. It had feathers and could fly like a bird, but had teeth and a bony tail like a dinosaur.

Layers of rock are stratified with older layers found below younger layers. Therefore, the fossil record makes for an excellent test for descent with modification. If species change over time, then we should find older species in lower geological layers, and modern species in more recent layers. If there is no descent with modification, then you should find fossils scattered randomly throughout geological layers. In this scenario of randomly placed fossils in geological layers, you could find the fossil of a rabbit next to a dinosaur. In every fossil bed in the world, we find that the fossils of ancient organisms coincide with the geological layer they should be in, fossils are not scattered randomly. Instead, the fossil record shows a clear progression in the evolution of life over millions of years. We never find a fossil of a rabbit next to a dinosaur. Therefore, the fossil record totally supports descent with modification!

Fossils and geological layers are aged using radioisotope dating. Remember that every element has isotopes, some of which are unstable and decay into other elements at a very constant rate known as the half-life. While some geological layers may be difficult to date, we can piece together the ages of most geological formations by knowing the ages above and below them. It's like putting together a large puzzle; it can be difficult to visualize where an individual piece may go. However, as you put more pieces together, the big picture becomes more clear making easier to understand where the individual pieces go.

Because the geological record is stratified in order of oldest rocks on the bottom and younger ones on top, geologists have pieced together an accurate picture of the age of many geological formations and fossils. This was put to the test in the early 2000s when a team of scientists led by Neil Shubin wished to discover a transitional organism linking the evolution of fish to terrestrial vertebrates. They had fossils of fish from 380 million years ago and potential early amphibians from 370 million years ago. Based on the dates of these fossils, they predicted that the ancestor to terrestrial vertebrates may have appeared about 375 million years ago.

To find such a transitional animal, they turned to geological maps to locate the right type of geological formations that would potentially have fossils. In 2004, they planned their trip to the Arctic Circle where they found the fossil they were looking for. It was named Tiktaalik and is a great example of a transitional organism with fish-like and amphibian-like characteristics. Finding Tiktaalik was yet another confirmation, along with thousands of other experimental and observational confirmations of evolution and descent with modification. Neil Shubin went on to write a book called *Your Inner Fish* about finding Tiktaalik and its implications for our understanding of vertebrate evolution.

Phylogenetic Trees Illustrate
Evolutionary Relationships

One of the remarkable insights of Darwin was that he realized that all modern species descended from ancestral species. Over time, species slowly changed due to relentless natural selection. Life diversifies over time as species evolve, adapting to their environment. Because modern species are the descendants of past species, they share a common ancestry and therefore share traits in common, which is the unity of life. The more recent a common ancestor between two species, the more closely related they will be to each other, and the more similar they will appear.

Let's use three of my favorite birds, the Rosy-Finches of North America, as a start to explain evolutionary relationships. These are small song birds that live above the tree-line in mountainous habitats of the western US. They are very closely related to each other sharing many features in common because they descended from a recent ancestor during the last ice age. The extensive glaciation separated the ancestor of modern Rosy-Finches into at least three different populations. With the populations isolated from each other and no longer able to interbreed, small changes began to accumulate in the populations, slowly making them different from each other. Once the ice ended, the populations were sufficiently different enough that they no longer interbred, even if they were to live together in the same areas again. There are now three species when historically there was only one.

Based on shared characters, we can construct a phylogenetic tree, which illustrates our best hypothesis of their evolutionary relationships. Rosy-Finches are related to other songbirds, such as the Northern Cardinal, but there have been millions of years of evolution separating a Northern Cardinal from a Rosy-Finch, so they share less in common with each other. All birds have feathers and a bill made of keratin that lack teeth, indicating there was one common ancestor about 135 million years ago that gave rise to all modern birds. Go back further in time, and birds and crocodiles shared a common ancestor some 240 million years

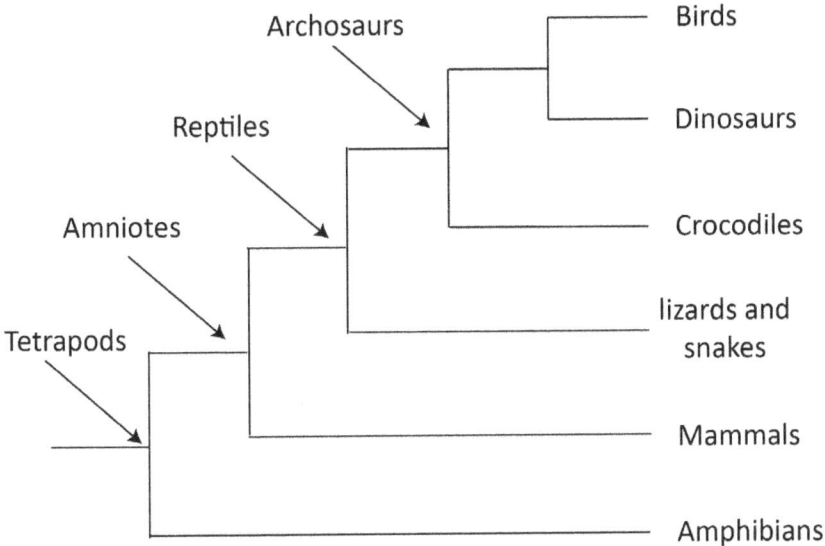

A phylogony representing the relationship among terrestrial vertebrates. The arrows point to a common ancestor for each lineage. For example, mammals and reptiles both amniotes, they have an amniotic sac and reproduce internally.

ago. In that time, birds and crocodilians have greatly diverged, yet they share common characteristics, including a four-chambered heart and other behaviors including making nest for their young, parental care, and they sing to their mates. Although, it's hard to call an alligator's deep guttural vocalizations a song.

Alligators and birds are living descendants of a much larger and once diverse group of reptiles called the Archosaurs (arch-ruler, saur – reptile, or ruling reptiles), which included the dinosaurs. Dinosaurs have been extinct for more than 65.5 million years, so it can be difficult to find clues to how they lived. Phylogenetic trees like the one illustrated on the next page can help us make predictions about the behavior and characteristics of extinct animals by comparing groups of living animals. In our bird and crocodile example, we know they share several behaviors, including making nest, parental care, and singing to their mates. Because dinosaurs were Archosaurs, we can predict that they most likely made nest, had parental care, and possibly sang to their mates. Two of these predictions have been verified with the finding of fossilized dinosaur nests with females still on top of their eggs. Birds and crocodiles both have four-chambered hearts, so we would hypothesize that dinosaurs also possessed four-chambered hearts, lending credence to modern theories that dinosaurs were active animals.

If you go back to the Pennsylvanian (part of the Carboniferous Period) some 320 million years ago, you would find the ancestor to both reptiles and mammals. It would have reproduced internally and had an amniotic sac, all adaptations to survive in a drier environment by reducing its dependence on having to return to water for reproduction. Going back another 50 million years, and you would find the ancestor to all tetrapods (tetra = 4, pod = foot) living near stream banks in vast swampy forests of the Devonian around 370 million years ago. To this day, the tetrapods include amphibians, reptiles, birds, and mammals, most of which possess four limbs (or have lost them as in snakes and whales). Not only do they possess four limbs, they also have the common bone pattern of one bone, two bones, lots of bones; a feature we inherited from that distant ancestor with both fish and tetrapod characteristics. Continuing our journey back in time to 500 million years ago to the Cambrian, you would find small "worm-like" animals that were the ancestor to vertebrates, which include all sharks, fish, amphibians, reptiles, birds and mammals. Similar to us, they had a dorsal nerve cord, a notochord, muscular tails, and pharyngeal slits modified into gills for breathing. These are the defining characteristics of vertebrates that set us apart from all other animals. In the tetrapods, the gill slits have been modified to form parts of the neck, jaws, and inner ear bones.

In the last 500 million years, those features have become greatly modified to accommodate the different life styles of vertebrates. If you were to go back even further, to 550 million years ago near the end of the Proterozoic, you would find primitive animals with true tissues, muscles, nervous system, and they were eating by ingestion. Today, all animals from spiders, to snails, to earth worms, to birds and mammals, all possess traits that were present in our most ancient animal ancestors. By comparing the similarities of all living animals, we can make certain predictions about our common ancestor. So far, all the evidence we have collected so far, including the fossil record, shared morphological traits, and the history written in our DNA, all tell the same evolutionary story. No other theory, idea, or ideology makes such powerful predictions of evolution that have been supported by repeated observations and tests.

Scientific Challenges to Darwin's Theory

Despite the predictive power of Darwin's work and the enormous amount of support he presented for the theory, there were several valid criticisms of the day. Most of the scientific criticisms stemmed from a general lack of knowledge about genetics and inheritance at the time of its publication. For example, no one knew how traits were passed from one generation to the next. The prevailing theory of the time was blending inheritance. Critics of natural selection argued that any variation would be swamped out over generations, so any positive change would not be able to accumulate in populations. Additionally, no one knew how variation in a population would be generated in the first place. Without a continual source of variation, evolution would eventually grind to a halt because there would be no variation to act on.

Some critics at the time also thought that the age of the Earth was much too young for all the species present today to have evolved slowly over time. In fact, the idea of a young Earth is still pervasive in society today despite enormous amounts of data to utterly refute it. But, in the 1860's there really was no way for geologists to accurately age the Earth, it took the discovery of radioactivity, the calculations of half-lives, and nearly a 100 years of work to accurately determine the age of the Earth. Much to the surprise of geologists and physicist working on the problem, multiple lines of evidence support the Earth to be 4.6 billion years old. An age that was far older than practically any one ever thought at the time.

Many critics continually point to the fossil record as being incomplete and that it lacks transitional species. It is true that there is not a complete fossil record of every species that has ever existed on the Earth. In fact, it is biased towards animals with hard parts that are easier to fossilize.

Despite these shortcomings, the fossil record fully supports evolution as explained in the previous section. Older fossils are found below younger fossils and they show a clear progression in age based on their position in geological layers. If you were to hike down the Grand Canyon, you would be hiking back in time, continually finding older fossils as you hike deeper into the canyon. Fossils are not randomly distributed in the geological strata.

Lastly, many people just simply did not accept evolution because it was contrary to their everyday experiences, it went against their common sense knowledge. However, the life of a human is a tiny fraction compared to geological time scales that span millions, or even billions of years of evolutionary history. Additionally, the *zeitgeist* of western civilization was influenced by Aristotle who believed that all species were created perfect and that any deviation would be a deviation from perfection. As a result, many in society rejected the idea that species change over time because they believed the world to be relatively young and unchanging.

Over the course of time, the weight of scientific evidence has completely disproven these criticisms. Evolution has become so well established that we know evolution is a fact. We can directly observe it occurring naturally or the laboratory; we can measure the rate of evolutionary change in organisms; and we can test for it to experimentally verify it. Darwin's theory of natural selection, which explains how evolution happens, has been repeatedly verified by thousands of independent observations and experiments. Today, evolution by natural selection is one of the most well supported theories in science, along with relativity, quantum mechanics, and the Big Bang.

Evolution by Natural Selection
The Theory that Revolutionized the World

Darwin's theory of evolution by natural selection caused a paradigm shift in the way we do science. Although Darwin wasn't the first naturalist to discover evolution by natural selection, in the 1840s, a forester named

Mathews devised practically the same theory as Darwin, but thought that evolution by natural selection was such common sense, he didn't publish it. Despite not publishing his theory, he thought he should have received credit for it. More than a decade later in the 1850s, Alfred Russel Wallace came up with evolution by natural selection and sent his theory to Darwin. Luckily for Darwin, he had a long, written history of developing his theory first, so he gets the majority of credit for evolution.

Evolution by natural selection was important for two reasons. First, it was an elegant theory that explained the unity and diversity of life. It fit nicely with the Linnaean classification of life and explained the patterns in the fossil record. It is also important for practical applications including cancer research, the emergence of antibiotic resistant diseases, conservation, and agricultural practices.

The second major contribution was that it fundamentally changed the philosophy of science by showing that scientific explanations can be used to explain all natural phenomena. Prior to Darwin's time, most naturalist, including Linnaeus, believed that all life was created basically in its present form at some fixed time in the past, a belief that was a hold-over from Aristotle. No one really searched for answers as to why there are so many species or their origins. If they did, supernatural explanations that were outside of the realm of science were often presented as an answer. By developing the theory of evolution by natural selection, Darwin demonstrated that there was no need for supernatural explanations to explain the natural world. Rather the answers could be determined based on scientific observations that are measureable and repeatable. Using a scientific explanation to explain the origin of species resulted in a profound paradigm shift that spread to all branches of science. It was an intellectual revolution paving the way for the most rapid expansion of scientific understanding in the history of our species.

Chapter 5

The History of Life is Written in Our Genes

There is more than 3.7 billion years of evolutionary history written in our genes

Introduction

The history of life is written in our genes, contained in the sequence of just four letters known as nucleotides. With today's technology, we are able to rapidly sequence entire genomes of organisms, shedding light on their evolutionary history and strongly corroborating the fossil record in support of evolution. Modern sequencing has allowed us to determine the relationships among modern organisms and even provide rough estimates of when they last shared a common ancestor with related species allowing us to construct evolutionary trees.

By comparing genes of numerous species, we have discovered that we share genes with every organism on this planet including plants, fungus, and even simple bacteria. These shared genes have been inherited from generation to generation dating back millions to billions of years ago. We also possess the remnants of genes that no longer work, but were once used by our ancestors millions of years ago. We also carry the genes of ancient viruses that have inserted their genes into our ancestors, hitchhiking along with every generation!

If only Darwin knew about heredity and genetics when he wrote On the Origin of Species, it would have added even more support to his theory and would have instantly silenced many critics. At the time of his writing in the 1850s, the prevailing hypothesis of heredity was that traits were blended, so any beneficial variation would be swamped out of over time. Adding to the problem was that no one knew how variations in traits could arise in the first place; as a result, Darwin's theory was vigorously debated for decades. By the 1940s, 80 years after Darwin published his theory, the evolution debate was largely settled as the New Synthesis emerged when Mendelian inheritance, modern genetics, and Darwin's theory of evolution by natural selection were combined into one comprehensive theory.

By the 1950s, it was known that at the heart of every cell is DNA, a molecule holding the genetic information necessary to make every living organism in the world. The discoveries of DNA's structure and the genetic code a decade later have been considered among the greatest scientific advancements of the 20th century. By making these discoveries, it linked heredity and evolution directly to molecular biology.

We are in the midst of a revolution in science as our understanding of how genomes evolve and work continues to grow at a rapid pace. Today, we can rapidly and cheaply sequence DNA, and the cost and speed of genetic sequencing continues to decline. Soon, you will likely have your genes sequenced as part of your medical history. With the advent of modern genetics, we can now create genetically modified organisms by inserting genes of one organism into other organisms. Even more recent technological breakthroughs are allowing us to edit genomes with incredible precision, which could have major implications for medical advancements and the development of genetically modified organisms (GMOs). To begin our understanding of heredity, I start with the pioneering works of Gregor Mendel and Thomas Hunt Morgan.

Heredity

For more than 3.7 billion years, the continuity of life has remained intact as the information in DNA has been faithfully replicated from one generation to the next. The implications of this statement is that you represent an unbroken lineage going back more than 3.7 billion years. This is the basis of heredity, the passing of genetic information from parents to offspring. However, we haven't always known how heredity works. When Darwin published his theory of evolution by natural selection, there were several major gaps in knowledge that were reflective of the times. First, it was not known how traits are passed from one generation to the next. Second, no one knew how variation in a population could arise. If you think about humans, what's the source of black hair, blond hair, or red hair? Fortunately, both of these questions have been answered with advancements in our understanding of genetics.

In the 1860s, evolution by natural selection was being debated among the naturalists of the time over how new or beneficial traits could arise and be passed to the next generation. It would take the pioneering work of Gregor Mendel and Thomas Hunt Morgan to shed light on the nature of inheritance and how it supports evolution. Gregor Mendel, known for his pea experiments, determined how traits are passed from one generation to the next. At the start of the 20th century, Thomas Hunt Morgan determined the source of variation in populations and showed that genes are located on chromosomes. By the 1940s, the theory of evolution by natural selection was combined with the work of Mendel and Morgan into a "modern synthesis", a term coined by the biologist Julian Huxley. You may recognize the name Huxley, Aldous Huxley was his brother who wrote the book, Brave New World published in 1932. Today, the modern synthesis remains the central paradigm, of biology.

Gregor Mendel – 1860s

The age of modern genetics begins in the 1860s with the studies of Gregor Mendel, but his work remained mostly unknown for 30 years until its rediscovery in the 1890s. Mendel experimented with pea plants to determine how traits are inherited (today we refer to traits as genes). In mid 1800s, it was generally assumed that traits were blended from one generation to the next, like mixing colors in paint. If there was variation, then it would be diluted over several generations, eventually being lost. This was one of the main criticisms of Darwin's theory. Mendel knew that organisms had traits that could be inherited, for example pea flowers were either purple or white and sometimes, traits would appear skip a generation. This pattern of inheritance didn't quite follow the main hypothesis of the time and there weren't many other testable hypotheses being proposed.

   To gain an understanding of how traits (genes) are inherited, Mendel performed numerous breeding experiments crossing pea plants with different traits (flower color, seed color, *etc.*) When Mendel crossed a true breeding purple flower with a true breeding white flower, all the offspring had purple flowers. Rather than stopping the experiment, Mendel continued by crossing the first generation (F1) of all purple flowers. In the second generation (F2), both traits were present; approximately 75% of the plants had purple flowers and approximately 25% had white flowers. The white flower trait, apparently "skipped" a generation. Fortunately, Mendel was a good mathematician and realized the significance of the 3:1 ratio of purple to white flowers in the second generation.

Mendel correctly interpreted the data from his experiment providing a much needed break-through in our understanding of inheritance. He realized that each parent contributed a gamete carrying a trait, such as flower color, to the next generation. He also knew that there were two types of variation for a trait, in the case of flowers, they were either purple or white. His next conclusion was that some traits were dominant and others were recessive. In this example, the purple flower color was dominant, it only required one copy of that trait to be expressed, whereas the white flowers were recessive. It took two copies of the trait to express the white flowers. That's why the first generation of pea plants in his experiment were all purple flowers, they all possessed one copy of the purple trait. When breeding the first generation with themselves approximately 25% of the offspring possessed two copies of the white trait. From this point on, I will use the modern terminology of genes instead traits, and alleles for variations in genes.

Today, Mendelian genetics refers to a specific scenario when on gene determines a characteristic such as flower color, which has two variations, known as alleles. In the pea flowers, there are two versions of the gene for flower color, a white allele and a purple allele. In Mendelian genetics, genes are also inherited independently of each other. For example, flower color has no influence on the inheritance of seed color or seed shape.

The results of Mendel's experiments are applicable to other sexually reproducing organisms including humans. In sexual reproduction, males and females produce gametes, each carrying a copy of their parent's genome; the gametes combine to form a new offspring. Therefore, sexually reproducing organisms carry two copies of their genome, one from each parent, meaning we are diploid. Humans have 46 chromosomes, or 23 pairs of homologous chromosomes. Homologous chromosomes are the same length and have the same genes in the same location on the chromosome. Because we have two copies of our genome, we can have two alleles that are the same, or two different alleles. If an organism possesses two different alleles for a gene, where one is dominant and the other is recessive, then organism won't show any sign of the recessive allele. This was the case for the first generation of pea plants that were all purple; the offspring had one allele for white flowers, and one allele for purple flowers. The second generation (F2) would have individuals with three combinations of alleles (two purple, one purple and one white, and two white) resulting in plants with purple and white flowers.

An important finding of Mendel's experiments was the discovery that genes are discrete entities, passed in whole from one generation to the next, clearly supporting particulate inheritance. Mendel's findings disproved the hypothesis of blending inheritance that plagued Darwin's theory of evolution by natural selection. Because genes are discrete units, they won't necessarily get "swamped out" in a population. In fact, favorable genes are more likely to persist and increase in frequency in a population over time. Whereas bad genes will tend to decline or even perish over time. Hence, species change over time as they accumulate favorable genes improving their "fit" to the environment.

To fully understand heredity of flower color in Mendel's pea plants, we need to apply our knowledge of modern genetics and chemistry, knowledge that wasn't present in Mendel's time. Pea plants with the purple allele produce a pigment called anthocyanin that builds up in the flower pedals turning them purple. Anthocyanin is the same pigment that gives blueberries and other fruits their dark blue color. At some point in the past, a mutation occurred in a pea plant, damaging the gene that produces anthocyanin resulting in a new allele producing white flowers.

In pea plants, they only require one working copy of the gene to produce anthocyanin, resulting in purple flowers. Because only one copy is required for the gene to be expressed, it is considered a dominant allele. To have a white flower, requires two copies of the white allele, so it is considered a recessive allele. It has nothing to do with one allele being more common or stronger than another allele. Luckily for that pea plant, pollinators also come white flowers, so the mutation causing the white allele wasn't detrimental and was kept and passed on to subsequent generations. Keep in mind that Mendel really didn't understand what a gene was or the source of alleles in a population, and neither did Darwin. That understanding would begin emerge 40-50 years later in the early 1900s with the pioneering work of Thomas Hunt Morgan and his gifted graduate students, some of who went on to win Nobel Prizes.

Mendel's experiments with pea plants was the first rigorous testing of inheritance. But, to be blunt, he also got lucky by picking genes that followed very simple rules of inheritance. This pattern does work for a very small number of genes. However, by the turn of the century, work by Thomas Hunt Morgan showed that patterns of inheritance are much more complicated than depicted by Mendel's experiment. Even though Mendel's view of inheritance was very simplistic and far from the complete picture, he was able to use his results to disprove blending inheritance by showing that genes are inherited as a discrete unit. His contributions to our understanding of heredity fully support Darwin's theory of evolution by natural selection.

Thomas Hunt Morgan and the Fly Lab - 1910s

While Mendel knew that traits were passed from parents to offspring, he didn't understand how or where they were located. Part of the answer began to emerge in the 1870s with the discovery of chromosomes and two types of cell division in eukaryotes, mitosis and meiosis. Mitosis produces two genetically identical offspring. Prior to mitosis the DNA is replicated and then condenses into visible chromosomes. It is during mitosis that replicated DNA is separated into two genetically identical daughter cells each receiving two copies of the genome. Almost every single cell in your body reproduces by mitosis. Because there is no recombining of genomes, mitosis is a form of asexual reproduction.

Meiosis was discovered in cells that form gametes, the eggs and sperm. It takes two rounds of cell division and results in 4 daughter cells that are genetically different. Each gamete carries only one copy of the genome, instead of two copies. The discovery of gametes, carrying only a single copy of the genome supported Mendel's hypothesis that a sexually reproducing organism receives a copy of each gene from one parent. Further experiments determined that chromosomes were made of proteins and DNA. Although, at the time, no one was sure of their structure, how they functioned, or even which of the two molecules was responsible for housing the genetic information.

The next major step in our understanding of genetics began in the early 1900s from studying fruit flies (*Drosophila melanogaster*). It's amazing that much of what we know about genetics has come from studying a small fly that thrives on ripe fruit. Fruit flies are what we call a model organism and have been a favorite of genetic studies for more than a century. First, they are easy to keep and rapidly breed in captivity. Second, they have 4 pairs (8 total) of large chromosomes, making them relatively easy to work with.

The importance of fruit flies to our understanding of genetics began with a single observation of eye color. The vast majority of fruit flies have red eyes, but one day, Morgan noticed a single individual with white eyes. He began to wonder what caused the rare white eye condition and whether or not it was heritable. To answer his question, he began breeding fruit flies by the tens of thousands in his lab, eventually a few mutant flies with white eyes randomly appeared seemingly out of nowhere. The discovery of the white-eyed mutants confirmed that mutations appear randomly and serve as a source of variation in populations. After additional breeding experiments with the white-eyed mutants, it became clear that chromosomes contained the genetic information and the pattern of inheritance for eye color didn't follow the simple rules of Mendelian inheritance, rather eye color was linked to the sex of the fly.

In addition to eye color, Morgan's work on fruit flies demonstrated that many genes are inherited together because they are located on the same chromosome. In Mendel's pea experiment, the genes for seed color, seed shape, and flower color are all inherited separately because those genes are located on separate chromosomes. But, this is not the norm for most genes. Think about it this way; you have 46 total chromosomes, 23 pairs of chromosomes with each set coming from your parents. On those 23 pairs of chromosomes there are about 20,000 genes. What this means is that hundreds to thousands of genes are located on the same chromosome so they are inherited together. Using humans as an example to show how genes are inherited together, the gene for the ability to see red and green is located on the X chromosome, which also contains genes that determine our gender. Therefore, color blindness is linked with gender because the gene is physically located on X chromosome that helps determine sex. If you're a color-blind male, then you inherited color blindness from your mom.

By the 1920s, Morgan's group had greatly expanded on the early work of Mendel. They discovered mutations are the source of variability in a population by producing new alleles, the chromosomal basis of inheritance (genes are located on chromosomes), that many genes are inherited together, and his student Hermann Muller created the first maps of genes on chromosomes. The work of Morgan and his graduate students was instrumental in bringing in the modern age of genetics and reconciling the scientific criticisms of natural selection with a modern theory of evolution. Ironically, early in his career, Morgan was a harsh critic of Darwin's theory of evolution by natural selection. However, after his pioneering work on fruit flies, he changed his mind once he realized the connection between heredity, genetics, and evolution.

Over time, our knowledge of genetics and heredity have greatly expanded and we now know of many other extensions to Mendel's work. For example, there can be more than two alleles for a gene. Keep in mind than an individual can only have two alleles, but there can be many alleles for a gene in a population. We can use the 'ABO' blood type as an example. Imagine the scenario where your mom has type 'o' blood, your dad is type 'AB', and you are 'A' blood. In this case there three possible alleles, but since 'o' blood is recessive, your mom would only donate the allele for 'o' blood. Your dad possesses two alleles, 'A' and 'B', so if you are blood type 'A' then you would have the recessive 'o' allele from your mom and the dominant 'A' allele from your dad. However, you could have also received the 'B' allele from your dad. As you can see, there are more than two alleles possible, 'A', 'B', and 'o'.

The 'ABO' blood type also offers another extension to Mendel's inheritance. Recall that in our imaginary scenario, your dad is blood type 'AB' meaning he is expressing both the 'A' allele and the 'B' allele. In this case, they are codominant meaning they are both expressed in an individual. Other alleles exhibit incomplete dominance because they appear blended. When a red snapdragon is crossed with a white snapdragon, the offspring all have pink flowers. In this case, it takes two functioning alleles to form a red flower. Only one functioning allele produces half the red pigment, so the flower is pink. Sometimes, there may be many genes working together for a particular trait like eye color where it has been estimated there may be up to 15 genes involved, explaining why there is so much variation in eye color among humans. In contrast, a single gene may effect numerous traits, like the sex-determining region of the y chromosome (SRY-gene), which turns on many other genes early in development and is responsible for turning a female fetus into a male.

The Importance of Mutations for Evolution

When we think of mutations, we often conjure up images of the super cool Teenage Mutant Ninja Turtles or the X-Men, or a three-eyed fish, or glow-in-the-dark animals. Other than Ninja turtles or X-Men, mutations are often considered to be bad and rightly so because some mutations, even very small ones, can be so bad they are lethal to the organism. However, they are not always bad, contrary to popular belief, mutations are very important because they are the ultimate source of genetic variation within a population. Mutations in genes within a population form new alleles as we saw in the purple and white pea flowers, the red and white eyes of fruit flies, or the ABO blood type in humans. Without variation, populations could not evolve because every organism would be identical and there would be no natural selection driving evolutionary adaptations. Clearly this is not the case as there is immense diversity in nature. Over billions of years, enough mutations have occurred that organisms as different as humans, fungi, plants, and bacteria have been able to evolve.

Despite the incredible accuracy of DNA replication, mutations, do occur. A mutation is any change to the sequence of DNA and can range from the very small to the very large. Small mutations would include the switching of one nucleotide for another, or the loss or addition of a single nucleotide. Imagine the word 'out', change the 'o' to a 'c' and you have a whole new word, 'cut'. Although these are relatively small mutations, their range of effects on the organism receiving them could be nothing to being lethal. Mutations can also be very large such as the duplication of small segments, genes, whole chromosomes, or even entire genomes.

In the big picture, mutations can be harmful, neutral, or even beneficial. Sometimes mutations can be harmful in one environment, yet be beneficial in others. A great example occurs in the skin color of our own species. The first humans evolved near the equator in Africa where 12 hours of sunlight occurs every day of the year and can be quite harsh on our skin, which unlike other animals is not protected by a fur coat. As our ancestors became less hairy, they evolved the ability to produce lots of the pigment melanin to protect their skin from the damaging effects of ultraviolet light. If you've ever gotten a tan, it's because your skin began to produce more melanin to protect itself from the sun.

Over thousands of years, humans migrated out of Africa and northward into Europe and Asia where there is much less sunlight, especially in the winter. Very dark skin in these latitudes is disadvantageous because it reduces Vitamin D production, which is not good for our health. In some populations, mutations occurred in the genes responsible for producing melanin in our skin, hair, and eyes, leading to people with light skin, blond hair, and blue eyes. In this case, mutations creating new alleles with reduced melanin production causing pale skin, allowing for more Vitamin D production, were beneficial and kept. Over time, these beneficial alleles accumulated in the population allowing our ancestors to adapt to the northern environments. On the other hand, if the same mutation causing pale skin occurred near the equator, it would be harmful and wouldn't last long in the population. As you can see, skin color is merely an adaptation in humans to protect us from sunlight, the mutation can be either beneficial or harmful depending on the environment.

Some mutations at first may be neutral, providing no immediate benefit to an organism so natural selection does act upon it. One example is a gene duplication, which leads to families of genes that increase the size of genomes and the number of genes. Once a gene is duplicated, it can mutate and natural selection can act upon the mutations to create a new gene with a novel function. Over time, additional genes allowed life to evolve and become increasingly complex as it has more genes to work with.

A great example of a gene family produced by gene duplications is reflected in the ability of mammals to detect odors. If you've ever been around the family dog, you've noticed their incessant sniffing of practically every object in their vicinity. The ability to detect odors is dependent on the number and types of odor receptors you have, which of course, is determined by genes. Early in the history of mammals, the genes responsible for detecting odors duplicated over a thousand times. Each time a mutation created an additional copy of the odor detecting it gene, it was subjected to mutations and natural selection. What entailed was the evolution of a gene family that allowed mammals to detect an incredibly wide range of odors.

Humans are not capable of detecting odors as well as a dog, even though we are mammals. But, we actually have the same genes or, more accurately, remnants of the genes that dogs have to detect odors! Instead of having nearly a thousand functioning genes, we have about 500 functioning genes producing proteins to detect odors. The rest are mutated and no longer function, so they are called pseudogenes.

Part of the answer to why we have lost our ability to detect odors lies in a trade off with improvements to our eyesight. Primates are better at detecting color than dogs and most other mammals because the genes used to make proteins to detect light and color were duplicated in the ancestor of primates perhaps 60 million years ago. This gene duplication is similar to the gene duplication that led to better sense of smell, except in this case, it led to a better sense of vision. The gene duplication coincided with the evolution of flowering plants and the nutritious fruits they produce. As a result, our diets are quite varied, think about all the colors of the fruits and vegetables we eat. Through mutations, our ancestor's eyesight improved while the necessity to detect numerous odors became less important. When a mutation would occur knocking out a gene required to detect a particular odor, it wasn't bad to our ancestor, so natural selection never acted upon it and the dysfunctional gene was passed on to the next generation rather than being eliminated from the population.

In an extreme example of relaxed natural selection, where mutations resulted in loss of function with little consequence, occurred in whales and dolphins. They have all the genes to detect odors, similar to a dog, but over evolutionary time, their genes to detect odors became mutated resulting in pseudogenes that are no longer functional. But, that's ok, because they don't need to detect odors in their environment to be successful. Natural selection did not remove those mutations from the population because they were of little or no consequence!

The Structure and Replication of DNA - 1950s

By the early 1950s, it was well-known that genes were located on chromosomes and experiments from the 1940s strongly indicated they were housed in the DNA and not the proteins. However, no one knew the structure of DNA or how it stored genetic information. As you may have guessed, the race was on to be the first to correctly determine the structure of DNA to determine how it stores information. Developed in the 1940s, a technique called X-ray crystallography has been used ever since to determine the shape of molecules by shooting high energy X-rays at a crystalline form of the molecule under study. Early on, this was a labor-intensive practice, often taking multiple attempts and sometimes several years to obtain one good image of a single molecule. Once a usable image was generated, it required expertise in chemistry and physics to correctly determine the shape of the molecule in question.

Unfortunately, the discovery of DNA is not without its controversy. Rosalind Franklin was an expert at X-ray crystallography, and it was her research lab at Kings College in London that had taken one of the best X-ray images of DNA. In 1953, the X-ray image known as Photo 51 was given to James Watson and Francis Crick without her knowledge. Using Rosalind's image, they correctly deduced the structure of DNA and quickly published the results in the prestigious journal *Nature*. Follow-up work for the rest of the 1950s supported their findings, firmly establishing the importance of DNA as the molecule of heredity. Less than a decade later, in 1962, Crick, Watson, and Wilkens each shared in the Nobel Prize for "their" discovery of the iconic double helical structure of DNA. Unfortunately, Rosalind Franklin was not awarded the Nobel Prize, she had passed away four years earlier from ovarian cancer.

DNA Structure

Discovering the structure of DNA has been hailed as one of the most important discoveries of the 20th century. Its iconic double helix has been practically synonymized with scientific advancement. DNA is a very large molecule made of just four building blocks called nucleotides (adenine, thymine, guanine, and cytosine). The length of DNA is determined by the number of nucleotides, which can be highly variable among different species.

DNA
Nucleotides
Guanine
Adenine
Cytosine
Thymine

An average bacteria, may have about 5 million base pairs, but the length can be over a thousand times longer in plants and animals. In humans, our DNA is about 3 billion base pairs if stretched out, it would about 6 feet long.

DNA is a actually a double-stranded molecule held together by weak hydrogen bonds between the four nucleotides. To visualize the structure of DNA, imagine taking a latter and twisting it to form a double-helix. The inside "rungs" are formed by weak chemical bonds between the nucleotides of the two strands. In any strand of DNA, the base pairing between the nucleotides is always the same: thymine always forms two hydrogen bonds with adenine (A=T) and guanine always forms three hydrogen bonds with cytosine (G≡C). As a result, both strands contain the same genetic information. The weak bonds between the two strands are easily broken and reformed, an important aspect of the double helix structure that allows it to be easily replicated or transcribed, a process I will discuss next. The outside part of the double helix, holding the "rungs" are formed by a sugar-phosphate backbone which is held together by stronger chemical bonds (covalent bonds for those interested in chemistry) that do not easily break.

DNA Replication

If there is to be continuity of life, then the information contained in DNA must be accurately replicated to form the next generation. Once the structure of DNA was discovered, Watson and Crick quickly deduced how DNA was replicated, which was quickly verified by experimental data. DNA replication requires the coordinated effort of at least nine proteins in addition to environmental ques triggering cells to replicate their DNA.

DNA replication begins when proteins separate the two strands of DNA, unwinding the double helix, allowing another protein called DNA polymerase to synthesize a new daughter strand of DNA from each of the parent strands. The copying of DNA is very precise, with only about one mistake in 100,000 base-pairs. After replication, there are two new daughter strands of DNA. The mode of replication is considered semiconservative because each new double helix is comprised of one original strand and one daughter strand. Other proteins are involved with proof reading the DNA, making sure that the information is correctly copied.

The DNA in our cells is not one long continuous strand, instead it is chopped up into multiple linear chromosomes where the ends of the DNA are free. The linear structure of our DNA has a major implication: DNA polymerase, the protein responsible for copying DNA, cannot fully copy a linear strand of DNA to the very end. Every time DNA is replicated in our cells, a small portion of the ends are lost, shortening the DNA with each generation. What this means is that our cells have a limited number of cell divisions. Eventually, after many rounds of replication, the DNA will shorten to the point that it damages the genes required for the cell to survive, causing it eventually die. To temporarily prevent a loss of information from the shortening of DNA, eukaryotes have a repeating unit of nucleotides called telomeres at the ends of their chromosomes. Each time DNA is replicated, our telomeres are shortened resulting in one less replication our cells can carry out.

If there is a rule of nature, one would be that there are no perfectly adapted organisms, every innovation comes at cost. The same goes for multicellular organisms including animals. One cost of being a multicellular organism is cancer, the uncontrollable growth of cells that causes death. We have more than 10 trillion cells in our body whose replication must be completely regulated. If just one of our cells becomes damaged and escapes our body's controls on replication, it can become cancerous. With the discovery of shorter telomeres in older eukaryotic cells, it has been hypothesized that linear DNA in eukaryotes is an advantageous adaptation for animals by limiting the number times our cells can replicate to prevent cancer. If cells have a limited number of replications, then cancerous cells would eventually die after a few rounds of replication. The trade-off here is that having linear DNA places an upper age limit on animals. The cells in your body only have so many replications before they will die

But, you may have noticed that we are not born old, somehow the biological clock is reset with every generation. One way to reset the clock takes place during gamete formation, animals produce a protein called telomerase that lengthens the telomeres at the ends of the cell's DNA. By the time you are born, the genes that make telomerase are permanently turned off in all your cells. Unfortunately, mutations in some cancerous cells result in turning the gene to make telomerase back on, allowing cancer cells to replicate forever by continually replacing their telomeres. As you can imagine, studying the regulation of telomerase is an active area of cancer and aging research. After all, wouldn't it be great to reset our cells and remain young for a number of years, or be able to stop cancer growth?

Gene Expression
From Genes to Proteins - 1960s

After the structure of DNA was determined in the 1950s, the next big discovery was to determine how DNA stored information and how that information was used to make proteins. Gene expression is the flow of information from our DNA to proteins. At its most basic level, information stored in DNA is transcribed to an intermediate called mRNA, and then the information is translated into proteins in a large molecular complex called a ribosome. Cells can fine tune both transcription and translation to rapidly respond to changes in their environment, or in multicellular organisms they can express unique combinations of genes so that cells can form different tissues such as skin or muscle. A gene has traditionally been defined as a sequence of DNA that codes for a product such as a protein. However, this simplistic notion has been challenged as we have learned that a large portion of our genomes code for short sequences of RNA that may be involved with gene regulation and are not coding for proteins.

The Genetic Code

Sometimes, the beauty of life lies in its simplicity. Recall that DNA stores genetic information in the sequence of just four different nucleotides:

adenine (A), guanine (G), cytosine (C), and thymine (T). Take a moment to reflect on the implications of this statement: All the information required to make every protein, in every living organism, in millions of species from simple bacteria to a human, is stored in the sequence of just four letters. The English language with all of its tens of thousands of words, and almost limitless stories uses 26 letters, but DNA, the universal language of life only uses 4 letters to make every protein and organism on the planet!

The code of life was cracked in the 1960s when scientists determined how the four letters of life, A T G C, form words known as codons. Codons are a sequence of 3 nucleotides that codes for a specific amino acid (recall that amino acids are the building blocks of proteins). There are 64 possible codons (4 x 4 x 4 = 64), but 3 are what we call stop codons, which tells a ribosome to stop making a protein. Recall that there are about 20 amino acids used by all life, which makes an average of three codons for each amino acid, making the genetic code redundant. Additionally, the same codon always codes for the same amino acid, so the genetic code is also unambiguous.

For the most part, all life uses the same 20 amino acids, meaning that when you look at the trees outside, they use the same genetic code as you do, and they use the same 20 amino acids to make all their proteins! Although, every protein is made using the same 20 amino acids, millions of proteins can be formed by varying the sequence of amino acids determined by a single, universal genetic code conserved across all life. The universality of the genetic code among all living organisms is another piece of evidence supporting the theory that all life descended from a common ancestor. If only Darwin had such information. It would have quickly ended many rancorous scientific debates that stemmed from a lack of genetic knowledge. Additionally, there are several medical and technological implications of a universal genetic code that I will discuss at the end of this chapter.

Many people have called the genetic code the blue print of life. However, this is not a good analogy; a blueprint is a precise set of directions to build some object such as a house or a cell phone. A much better analogy is to think of the genetic code as your grandmother's cook book containing many recipes that work together to create an awesomely huge Thanksgiving dinner. And if you've ever used a cook book, you may have made small notes on the pages to slightly modify the recipe. What's interesting about gene expression, is that the environment can alter the activity of genes, similar to making notes in a cookbook to slightly alter the recipe. In the last few years, we have discovered that these small changes are not mutations to the DNA, yet they can be passed from one generation to the next. This exciting discovery is called epigenetic inheritance, a topic I will discuss in the next section.

Gene Expression is the Flow of Genetic Information

DNA does not directly make proteins, it is used for the long-term storage of information, much like a library stores information in books. You may check out a recipe book explaining how to make your favorite pizza, but the book does not make the pizza itself. Gene expression is the flow of genetic information from DNA to mRNA to proteins. The information in genes is first transcribed from a DNA template to messenger RNA (mRNA). Transcription is similar to copying notes from your text book or from a lecture. Once mRNA has been made, the information is then translated into a protein in a large molecular complex called a ribosome. Every protein on the planet was synthesized in a ribosome using mRNA. It's important to point out that DNA is not turned into mRNA and mRNA is not turned into proteins.

Early in the 1960s, it was thought that a gene is a sequence of DNA coding for a specific protein. In recent years, genetics has undergone an incredibly rapid growth of knowledge that has somewhat muddled the exact definition of a gene. Recent findings have shown that a single gene may be able to make several proteins. Also, some sequences of DNA do not code for proteins at all, instead they code for sequences of RNA, an extremely versatile molecule used to carry genetic information, help make proteins, and regulate gene expression.

Cells Regulate Gene Expression

You are not what you were 6 months ago. Seriously, almost every molecule in your body is turned over every 6 months. Yet, here you are in almost the exact same form as you were 6 months ago, including the same memories, body size, skin color, *etc.* Your body could be compared to a river, the river is always present, but the water is constantly moving through, just like molecules continuously move through your body. How does your body turn over almost every molecule within months, but you remain the same? The answer lies in how our cells regulate gene expression maintaining our bodies for decades.

In humans, we are made of more than 10 trillion cells that form about 200 different types of tissues from skin cells to nerve cells. The activity of genes in each cell has to be precisely regulated because cells do not need to be constantly expressing all their genes at the same time. Not only would it be terribly inefficient, but your cells couldn't differentiate into different tissues and organs making it impossible for us, or any type of plant or animal to exist. Each cell type expresses different proteins at different times, this explains why skin cells look so different from a muscle cell. Understanding how genes are regulated is a rapidly growing field where scientists are currently making new discoveries almost daily.

The ability of cells to regulate gene expression is also important so they can rapidly respond to changes in their environment and reduce wasted energy expenditures. Recall that every living organism must acquire energy and resources from their environment. Although individual organisms cannot adapt to their environment, they can quickly react to changes by fine tuning the expression of their genes to produce certain proteins when needed and turning off the production of others when they are not needed. Much like a volume knob on a radio, genes can be completely turned off, or ramped up to produce proteins at full volume.

To understand how rapidly cells can respond to their environment, imagine the billions of bacteria in your gut, they are dependent on the foods you eat for their survival. Because we sometimes drink milk, those bacteria have genes to produce protein enzymes that can break down lactose, a sugar found in milk. If you drink milk, then those bacteria will begin to produce enzymes that break down lactose. To conserve energy, they will stop producing the necessary enzymes when lactose is not present by turning off the gene expression by stopping transcription.

To control when the proteins required to break-down lactose are made, bacteria use an operon, which is a segment of DNA that contains the genes to make a protein along with the genes required to regulate the activity of that gene. Operons allow prokaryotic cells to rapidly respond to changing environmental conditions by ramping the activity of genes either up or down, like a volume knob. There are numerous operons in bacterial cells switching on genes and turning them off in response to changing environmental conditions.

One major difference between prokaryote and eukaryote genomes is size, eukaryotic genomes can be hundreds to thousands of times larger than prokaryotic genomes. Prokaryotic genomes are optimized for quick reproduction and have very little DNA that is potentially non-functional, whereas eukaryotes may have upwards of 80% of their genome that appears to not really be doing anything. Researchers are currently investigating why eukaryotic genomes are so large, whereas only a small portion, less than 5% of the human genome may actually code for proteins, and that is likely an over estimate.

At the start of the human genome project in the 1990s, scientists set out to sequence the entire human genome, a monumental task for the time that was projected to take over two decades and costs several billion dollars. By the end of the decade, the human genome was sequenced for a fraction of the originally estimated cost. When sequencing began on the human genome, friendly bets were made regarding the number of genes that would be present. Original estimates were placed around 100,000 genes; after careful analysis of the human genome, that number has been revised downward to be about 20,000 genes! Having so few genes has puzzled scientists and is a remarkable achievement of evolution considering that there could upwards of a million different proteins produced by our cells.

Scientific discoveries almost always lead to more questions and then new discoveries. Once it became clear there were far fewer genes in our genome than predicted based on the number of proteins we potentially make, scientists began to ask how does our genome make so many proteins with fewer genes? By carefully analyzing the genes, another key difference was discovered between prokaryotes and eukaryotes. They both use the same genetic code, but eukaryotic genes have regions of intervening units called introns embedded within a gene. The part that actually codes for a protein is called an exon.

You can think of it like watching your favorite movie (the exons) interrupted by commercials (the introns). The purpose or origins of introns is not yet quite clear. However, it turns out that a gene can have numerous introns that are removed prior to translation, and there can be multiple exons within a gene. In a process called alternative mRNA splicing, cells can rearrange the exons to make numerous proteins from a single gene! In some ways our protein coding genes are similar to YouTube videos where people splice movies and television shows together into new stories by using old content. Just watch Star Wars versus Star Trek or Darth Vader vs Batman where unique stories are created by splicing together different content.

Recent evidence suggests that some of the non-protein-coding segments of our genome may be used to make many types of RNAs whose primary function is to fine tune gene regulation. They can carefully control how many proteins are made and how long those proteins remain active inside cells before they become targeted for degradation. Evidence is beginning to indicate that the ability to fine tune the regulation of genes is part of the reason why eukaryotes have been able to evolve into complex animals.

Believe it or not, the eukaryotic story of regulating genes gets even more complicated. In the last decade, the progression of technology has allowed us to understand how genes work with even finer detail. We know that cells possess numerous mechanisms to regulate the expression of their genes including operons, alternative mRNA splicing, and the activity of small non-protein coding RNAs. However, it turns out that gene expression can be regulated by a process known as epigenetics, a word that literally means "above genetics". Basically, small molecules called methyl groups can be added to the DNA where certain genes are located, effectively turning off that gene by preventing its expression. If you've ever seen a calico cat, the pattern of orange and black fur is the result of genes being permanently turned off. The same occurs in our cells, for example our skin cells do not produce the proteins that our eyes produce to detect light.

We are just now learning that when our cells respond to the environment, those responses can be passed to our offspring through epigenetic inheritance. The first evidence of this came from the Dutch Hunger Winter in World War II when the Germans cut off food supplies to the Dutch. As a result, women that were pregnant prior to the famine and gave birth during the famine had smaller children. This makes sense because babies gain their most weight in the last three months of pregnancy. However, these children remained small throughout their lives. Even more remarkable, the grandchildren were also small. It turns out that epigenetic inheritance in response to starvation was passed down at least two generations. It is observations like these that are important so we can ask which genes are being affected by the Mom's environment, and how those patterns of epigenetic regulation are passed from parent to offspring.

Modern Advances in Genetics - 2000s

Modern Genetics began with Gregor Mendel in the 1960s and was greatly expanded in the early 1900s by Thomas Hunt Morgan. By the 1940s, the modern synthesis combined what was known about genetics with evolution by natural selection to create the paradigm of modern biology. One famous biologist, Theodosius Dobzhanski, once proclaimed that "nothing in biology makes sense except in the light of evolution". In the 1950s work by Rosalind Franklin, James Watson, and Francis Crick determined the structure of DNA and by the 1960s, the genetic code was cracked revealing the language of life. By the end of the second millennium, we had sequenced the entire 3 billion base pairs that form the human genome. Today, we can sequence a genome in a few days for less than $2000, and the costs continue to decrease.

The history of life is written in the genomes of living organisms. To understand that history, we have sequenced the genomes of bacteria, birds, mammals, fish, worms, and reptiles along with sequencing partial genomes of thousands of other organisms. By using sequences from many species, we have been able to reconstruct the relationships of many living organisms, understand how natural selection has altered species over time, and develop models characterizing ancestral species. We have also learned how genes are regulated in cells, quickly responding to environmental changes, or how other genes known as master regulatory genes, regulate the activity of numerous other genes during growth and development. By understanding how these genes work we can learn how one cell grows into an adult as our cells differentiate into different tissues, or how we grow limbs and organs. One of the goals of science is to figure out how we can regrow damaged tissues, organs, or even lost limbs.

Modern genetics is also rapidly revolutionizing medicine as never before. By understanding the workings of gene regulation, we are making huge strides in fighting cancer. We are also learning what causes our cells to age and die and what we may do to slow down or even reverse the aging process. Advances in stem cell research may someday allow us to regrow damaged tissues, organs, or even new limbs that have been lost or damaged.

One major implication of the universality of the genetic code is that we can engineer organisms by inserting genes from one species into another, making genetically modified organisms (GMOs). While the use of GMOs has become controversial among the public for various reasons, it has helped save the lives of thousands of people by allowing us to cheaply make drugs to cure diseases. One of the best examples of GMOs saving lives is the cure for diabetes when bacteria engineered with human genes makes insulin to lower thier blood sugar levels. People with diabetes often can't make insulin, so they require a constant supply of insulin or they could die.

In addition to engineering bacteria, scientists have also engineered many of our crops to improve yield and nutrition. Golden Rice was modified to produce beta carotene; a vitamin we require because our bodies can't make it. It's especially important for children because if they become severely deficient, it can lead to blindness or even death. Golden Rice could prevent thousands of cases of childhood blindness in the poorest regions of the world. Unfortunately, the countries that could most benefit by it have banned its import, erroneously believing it is bad or goes against nature. However, it's important to know that the beta carotene produced by Golden Rice is the same beta carotene produced by any other plant, including carrots. Recently, a new type of salmon has been created by inserting the gene from another closely related species allowing the GMO salmon to grow year-round, thus reaching a larger size more quickly. This engineered fish is not bad for your health and does not possess any unknown bad health effects. In fact, most GMOs have undergone rigorous testing clearly demonstrating that they are safe for consumption.

Scientific advancements in genetics is moving more quickly than society's ability to understand the implications of its findings. Therefore, it's important as a citizen to be informed about the implications of scientific advancement and make good, well-informed decisions based on facts and data.

Chapter 6

Reproduction
Ensuring the Continuity of
Life

You are the end result of an unbroken lineage going back
more than 3.7 billion years to the dawn of life

Introduction

You are the end result of an unbroken lineage going back more than 3.7 billion years to the dawn of life. All the life we see today, including ourselves, can be traced back to an ancient common ancestor that was most likely a small population of bacteria living in the ancient oceans. For some, it can be hard to imagine that a small cell using a natural energy source to synthesize organic molecules from carbon dioxide and hydrogen gas would eventually give rise to all the life we see today. Luckily for us, those primitive cells found a way to live independently of their place of origins and reproduce themselves. Reproduction ensures that life continues on, slowly evolving into the myriad of species we see today. I have often wondered how many times life may have emerged in the first billion years of our planet, only to perish because it never acquired the ability to reproduce or leave its place of origins.

Reproduction can be relatively simple such as a prokaryote copying its DNA and dividing into two genetically identical offspring, or more complicated, such as a eukaryotic cell reproducing sexually by combining two genomes to form one offspring. The evolution of sexual reproduction and two sexes presented a paradox perplexing biologists for over a century. Quite simply, why would you combine two cells to form one offspring? And why cut in half the number of potential mates in a population with the evolution of separate sexes? Obviously, there are advantages to sexual reproduction, but teasing out its evolutionary origins has been difficult and filled with lively debates. I present a few of the theories in this chapter, some controversial, others more widely accepted.

With sexual reproduction comes sexual selection, a force of nature that has led to colorful ornamentation, elaborate dances, intricate songs, or enlarged adipose tissue surrounding mammary glands, all to attract choosy mates. Indeed, the need for reproduction has driven the evolution of some very strange looking animals and odd behavior.

Prokaryotes Reproduce by Binary Fission

The first cells on the planet reproduced by binary fission a form of asexual reproduction. They copied their DNA, divided their cells roughly in half, and produced two genetically identical daughter cells. Sometimes, mutations would occur when they copied their DNA. Natural selection acted upon this variation allowing evolution to proceed in its random walk to an unknown future. Asexual reproduction produces genetically identical cells, which is a great way to reproduce as long as your environment remains relatively stable. You reproduce offspring that are fit to the same environment, over and over again.

Natural selection keeps prokaryotic genomes small. These small cells are adapted for rapid reproduction when times are good. Copying extra DNA cost the cell energy, time, and precious resources. Therefore, any unused genes, or extraneous segments of DNA, not serving a purpose are often lost after a few generations, mostly through random mutations, maintaining a streamlined genome primed for rapid reproduction. Prokaryotes never combine two whole genomes into one organism, but they can exchange DNA between cells, even when they are not closely related.

Let's take the well-studied *E. coli* bacteria found inside your gut as an example of a typical bacterial genome. These tiny bacteria come in different strains where the number of genes range from 4,377 to 5,416. The vast majority of these genes are involved in proteins synthesis to maintain the functioning of the cell. But, you get the picture here, the average size of their genome is about 5,000 genes, give or take a few hundred genes. This translates to about 5 million base pairs of nucleotides. Natural selection constantly trims their genome to maintain rapid reproduction.

Within your gut, there are billions of these tiny bacteria making a living off the foods you eat. They aren't parasites by any means, they are a vital component of our guts, aiding in digestion, producing beneficial molecules, and perhaps even preventing pathogens from causing us harm. Under favorable conditions, their population can double every few hours, thanks to small genome size and rapid reproduction. Although, we refer to *E. coli* as a single species, a considerable amount of genetic variation exists, where as few as 20% of the genes are shared. When you think about it, two bacteria we call *E. coli* are more different than you are to a tree! But, what they all have in common are similar metabolic pathways and genes required to let them thrive in your gut.

Prokaryotes Exchange Genes Through Horizontal Gene Transfer

"It's complicated", one of my favorite responses to what seems like a simple question. The evolution of life over the past 3.7 billion years leading to modern organisms, is "complicated". In the two previous chapters, I wrote that modern species evolved by descent with modification; mutations create variation in our genes, which in turn are selected for or against by natural selection. Over time, species change, reflecting changes in their genomes as they become better fit to their environment. But some genetic changes, including the evolution of new alleles or genes, may not have been created by a mutation in a parent and passed to their offspring. Sometimes, new genes are acquired from the environment and incorporated into the organism's genome. Prokaryotes have the ability to take in environmental DNA, random fragments of DNA floating around in their environment, which allows populations of prokaryotes the ability to rapidly evolve to changing environments.

With the advent of modern sequencing techniques, it has been found that the pangenome of *E. coli* may include some 16,000 genes, a number that rivals the human genome. The pangenome of a population includes all the unique genes in a single population of one type of bacteria. When bacteria reproduce asexually, this is called vertical gene transfer where genes are passed directly from parents to genetically identical offspring. But, prokaryotes also have the ability to take up genes from the environment, a process known as horizontal (or lateral) gene transfer. In horizontal gene transfer, bacteria can acquire segments of DNA from the environment carrying new genes into their own genome. These short sequences of DNA may originate from closely or distantly related organisms, they don't even have to originate from other bacterial cells. This swapping of genes may not be reproduction in the strictest form, but it illustrates how living organisms can operate in vastly different ways than what we are accustomed to in our own lives.

Horizontal gene transfer has also been impicated in the spread of antibiotic resistant diseases. Many types of organisms, including fungus and other bacteria, naturally have genes to produce antibiotics to protect themselves from bacteria. Those genes can enter into harmful bacteria through horizontal gene transfer and become highly selected for when antibiotics become common in the environment.

Prokaryotes can reproduce rapidly. It may be hard to believe, but there have been more generations of *E. coli* in your gut than there have been generations pf humans since we last shared a common ancestor with chimpanzees about 5 million years ago! The combination of rapid reproduction, a large population, a large pangenome, and the ability of prokaryotes to incorporate environmental DNA, bacteria populations are able to adapt quickly to their surroundings when environmental conditions change. To complicate our concept of species and evolution even further, recent studies indicate that approximately two-thirds of the *E. coli* pangenome may have originated from other bacteria!

Horizontal gene transfer is possible because of the universality of the genetic code. It also allows us to manipulate the genes of any organism, including bacteria to create beneficial strains for research and medicine. It also complicates our simplistic view of descent with modification, because prokaryotes acquire beneficial genes from other bacteria without inheriting them from their parent cell in vertical gene transfer. Therefore, when we talk about our last universal common ancestor (LUCA), we almost always mean a population of prokaryotes rather than a single individual.

What exactly are we? A seemingly simple question with a complicated answer. It turns out that prokaryotes are not the only organisms that have a genome with foreign genes, the result of horizontal gene transfer. Eukaryotic genomes also possess bacterial and viral genes, and lots of them, starting from the wholesale transfer of mitochondrial genes that took place in the early evolution of eukaryotes. Today, human mitochondrial genomes have retained about 37 genes, just enough to regulate important protein activity for ATP production. That places about 500 genes in our nuclear DNA from the ancestor of mitochondria. Other genetic studies have found nearly 100,000 viral fragments accounting for nearly five to eight percent of our genome. These ancient viruses inserted their genes into the genomes of our ancestors, only to be caught there and passed down through the generations. Luckily for us, mutations have rendered many of them harmless.

The origins of mammals may have been directly caused by horizontal gene transfer from a virus! A unique feature of mammals occurs in development when the placenta embeds itself into the uterus of the mother. A special protein called syncytin makes this connection happen. Once researchers sequenced the gene for this protein, they immediately noticed that it matched the genes found in a virus. What this means is that a very mammalian trait, the placenta, has a viral origin, at least in part. It appears that horizontal gene transfer may have directly helped lead to the evolution of modern mammals.

Eukaryotes Reproduce Asexually by Mitosis

The process of dividing the nuclear material in eukaryotic cells and forming two genetically identical daughter cells is called mitosis. Mitosis is an incredibly complicated set of coordinated actions carried out by the cell where everything must go exactly right, or the cell will die or even kill itself if it fails to separate the chromosomes correctly.

Eukaryotes may be known for the evolution of sexual reproduction, but prior to that evolutionary innovation, they reproduced asexually. I imagine at first, the process was similar to binary fission in bacteria; the DNA was replicated and then the cell divided into two genetically identical daughter

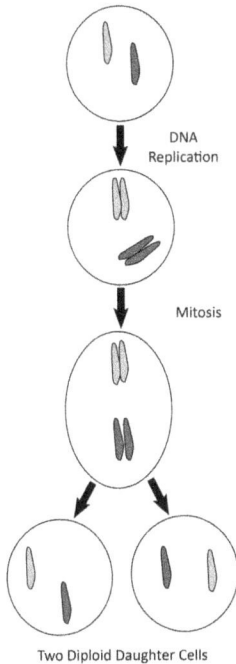

DNA Replication

Mitosis

Two Diploid Daughter Cells

Mitosis is asexual reproduction in eukaryotes. This cell has two copies of a single chromsome.

cells. We know that with the acquisition of mitochondria, ATP production was significantly increased, allowing eukaryotes to become much larger as they evolved internal membranes known as organelles to compartmentalize the functions of their cells. Several important changes took place in the DNA of eukaryotic cells. First, their DNA became associated with numerous proteins to help organize and regulate it as it grew larger. Second, it was cut into multiple linear segments to form distinct chromosomes unlike the single, circular chromosome found in prokaryotes.

When eukaryotes reproduce asexually, the DNA is replicated in a similar manner to prokaryotes. However, once the DNA is replicated in eukaryotes, it condenses into visible chromosomes that can be seen when dyed under a light microscope. Technically, prior to cell division, each copy of the chromosome is called a chromatid with both copies forming sister chromatids held together by glue-like proteins. You can think of sister chromatids as two copies of a single chromosome, which are comprised of DNA organized by numerous proteins. As we learned in the last chapter, the same segments of DNA always end up in the exact same place every time DNA is condensed into chromosomes. It is this order that allowed early geneticists like Thomas Hunt Morgan to create genetic maps illustrating the locations of certain genes on chromosomes. What this means is that each of our 23,000 genes has a unique place in one of our 23 pairs of chromosomes.

Once the DNA is replicated and tightly packed into chromosomes, the cell uses numerous proteins to divide the cell in half. While there are many important proteins required for mitosis, I'm going to focus on microtubules, these are a type protein that forms part of the cytoskeleton of cells. Similar to our skeleton, a cell's cytoskeleton provides support and helps to maintain its shape. The microtubules resemble hollow tubes, like a pipe, and are used by the cell to organize the chromosomes during cell division. If the cell is able to successfully move the chromosomes to the center of the cell, a chemical signal is sent cleaving the proteins holding the sister chromatids together. At this point, the replicated chromosomes begin to move to opposite ends of the cell as they are pulled apart by motor proteins moving along the microtubules. When the chromosomes reach the opposite ends of the cell, the microtubules dissemble and chromosomes begin to unwind from their condensed state. The last step involves pinching the cellular membrane in half forming two genetically identical daughter cells. We have spent over 100 years studying how the chromosomes are organized and pulled apart by the microtubules and understanding their importance during mitosis remains an active area of research.

Understanding how mitosis works has enormous implications for our growth, development, and health. All multicellular organisms are made of eukaryotic cells, and with the exception of our gametes, all the cells in our body reproduce by mitosis. Every day, you lose approximately a billion cells that must be replaced by mitosis. Recall from the last chapter that the activities of each one of these cells must be regulated, expressing only certain genes and reproducing only when needed. By learning how cells regulate mitosis, we may learn how to get new cells to grow when our tissues or organs become damaged. Imagine if we could grow new limbs for amputees or get our heart muscles to grow new tissue after a heart attack.

A price we pay for being a multicellular organism is cancer. When our cells lose their identity and begin to divide uncontrollably, they become cancerous. By understanding mitosis, we can understand how to fight cancer. You may have witnessed someone with cancer on chemotherapy lose all their hair and their skin looks very sickly. The reason is because many cancer drugs work by halting mitosis. For example, the cancer drug Taxol prevents the separation of the sister chromatids during mitosis. If cells become stuck in mitosis, they will actually commit suicide through a process called apoptosis. An unfortunate side effect of chemotherapy is that it prevents almost all of your cells from dividing, that is why cancer patients have sickly looking skin, hair loss, and often feel nauseous.

The Evolution of Sexual Reproduction

Sexual reproduction is a uniquely eukaryotic process where two copies of a genome are combined to form a unique organism. Genetic recombination is the hallmark of sexual reproduction; however, its evolution and persistence among eukaryotes has puzzled scientists for over a century. First, common sense would indicate that asexually reproducing organisms should outcompete sexually reproducing organisms because asexual organisms can give rise to their own offspring without having to find a mate. Second, asexually reproducing organisms should be able to produce offspring more quickly than sexually reproducing organisms. And third, another apparent disadvantage of sexual reproduction is that any parent would only be 50% related to its offspring rather than 100% related. Sexual reproduction can also break up favorable combinations of alleles in an individual, potentially producing slightly less fit offspring. However, within the past decade, new research has shed light on the age old question of the origins of sexual reproduction.

The mitochondrial theory for the origins of sexual reproduction implicates endosymbiosis and the origins of our good friend the mitochondria as an evolutionary driver of sexual reproduction. Imagine a scenario where you are living inside your host, which you have become totally dependent upon as you've begun to lose most of your genes to your host. But, that's not a problem, the host provides you with all the raw materials you require. In return, you produce copious amounts of ATP, that all important energy currency for the cell. Everything is going well, except that by using oxygen to produce lots of ATP, you also generate damaging free radicals that damage your host's membranes, proteins, and DNA. Free radicals are molecules that have unpaired electrons and are thus highly reactive. They want to quickly form bonds with any nearby molecule.

In response to more free radicals, it is thought that eukaryotes evolved a nuclear envelope to house and protect the DNA. But alas, the cell's DNA can still be damaged. If enough damage occurs, the host cell dies, taking the mitochondria with it. If your host only has one copy of its genome and it becomes severely damaged, that is not good for you or your host. Now, imagine if you could somehow bring in a second copy of a genome? If it's not an identical copy from a close relative, then it would be less likely to have the same damaged genes. Also, with two copies a genome, DNA repair mechanisms could fix the damaged DNA on one copy by using the other as template, guiding the repairs.

Recall that another paradox of sexual reproduction is that it can break up favorable combinations of alleles, or even produce unfavorable combinations. This problem was brought up when a famous comedian was once propositioned by a beautiful movie star suggesting that they should have children together. She told him that their children could have his brains and her good looks. But, the comedian replied, "They could also have my looks and your brains", a very possible outcome of sexual reproduction!

A second theory proposes that sexual reproduction can rid populations of deleterious alleles. It has been argued by some scientists that random mutations will eventually cause slightly bad or deleterious alleles can build up in a population. A single or even a couple of these slightly deleterious alleles may not be enough to prevent an organism from surviving and reproducing. In a population, these deleterious alleles may be spread across many individuals affecting the entire population. In sexually reproducing organisms, genetic recombination could put several of these deleterious alleles into one unlucky individual, basically rendering it unable to reproduce. In this case, the population would instantly rid itself of these undesirable alleles, thus improving the overall fitness of the population.

Admittedly, this theory has received some harsh criticisms because it implies that sexual reproduction may be better for a population rather than providing direct advantages to the individual. However, there is experimental data on some organisms to support this theory. But, most experiments on evolution support that natural selection operates on the individual, and it can be difficult to find evidence supporting natural selection acting on populations.

A third theory predicts that the advantage of sexual reproduction is that it increases genetic variation among offspring improving the chances of reproduction, especially in a variable environment. Sexual reproduction is not a source of new alleles, only mutations or horizontal gene transfer can serve as source of new alleles or genes in populations. But, sexual reproduction does create novel combinations of alleles that may work well together, the opposite of removing deleterious alleles from a population.

In the end, the evolution of sexual reproduction may have been driven by a combination of several factors where each improves the chances of reproductive success in the individual and improve the fitness of the overall population. However, to combine entire genomes into a new organism required a new way to reproduce. A recurrent theme in evolution is that it rarely re-invents the wheel, rather it modifies what's already there. In this case, sexual reproduction evolved by modifying mitosis.

Meiosis Increases Genetic Diversity

Sexual reproduction produces a new organism by fusing a male gamete (sperm) and female gamete (egg) and then combining their DNA together. If gametes were created by mitosis, then the number of chromosomes would double each generation. Therefore, meiosis evolved to produce gametes with only one copy of the genome. Recall that the eukaryotic genome is chopped up into multiple linear segments called chromosomes with each cell carrying two copies of the genome. As humans, we have 46 total chromosomes, 23 pairs that we inherit from each of our parents.

Meiosis almost certainly evolved from mitosis with just a few small changes. Meiosis generates four gametes that are genetically different by using two rounds of cell division. The term ploidy refers to chromosome number. Eukaryotes are typically diploid (di=two), meaning they have two copies of their genome. In humans, we have 46 chromosomes in all the cells of our body. Whereas haploid means one copy, so our gametes possess 23 chromosomes. Our chromosomes are numbered from 1 to 22 based on size with the X and Y chromosomes being the sex chromosomes. Therefore, you have two number 1 chromosomes, which are homologous chromosomes containing the same genes, but potentially different versions of those genes (alleles).

Prior to cell division in any eukaryote, the DNA is replicated and condenses into chromosomes. In mitosis, there are actually two copies of each chromosome held together forming sister chromatids, which are pulled apart during mitosis. But, meiosis does something a little different. Prior to the first cell division, the sister chromatids of homologous chromosomes join together to form a single large structure known as tetrad because it actually has four chromosomes.

What happens in the tetrad is very important for inheritance because it generates new combinations of alleles not seen in either parent. Recall that Thomas Hunt Morgan demonstrated that many genes are inherited together because they are physically located on the same chromosome. In humans, a great example of linked genes is red hair, green eyes, and fair skin. However, a process called crossing over occurs prior to the first cell division when sister chromatids of the homologous chromosomes fuse together into a tetrad. It is here that different alleles from your parents can be swapped, forming new combinations on a chromosome. This can occur because homologous chromosomes carry the same genes in the same location, but within a population, there may be different versions of those genes called alleles. The tetrad allows alleles found on your Mom's chromosomes to jump to your Dad's chromosome and vice versa, creating new combinations on a chromosome. Crossing over may lead to a person with red hair with dark eyes and dark skin, although this is rare.

Meiosis also generates new combinations of alleles by a process called random assortment of the homologous chromosomes. During the first round of cell division in meiosis, the tetrads line up in the middle of the cell and are pulled apart, similar to mitosis. When the tetrads line up in the middle, their orientation is random, meaning the daughter cells receive a random assortment of maternal and paternal chromosomes. The number of possible combinations of maternal and paternal chromosomes grows exponentially 2^n, where n=number of chromosome pairs. If an organism has a total of four chromosomes (2 pairs), then the number of combinations is 2^2 (2 X 2) = 4, if there are three pairs then the number of combinations is 2^3 (2 X 2 X 2) = 8. In humans the number is much higher, 2^{23}, which is just shy of 8.4 million possible combinations.

After the first round of cell division in meiosis, the homologous chromosomes have been separated, but the sister chromatids remain together. The second round of meiosis separates the sister chromatids, forming a total of 4 haploid daughter cells that become our gametes. In the act of sexual reproduction, two gametes randomly fuse forming a zygote, which grows into a new organism. In humans, there are approximately 8.4 million possible combinations of paternal and maternal chromosomes in each gamete, when they fuse that creates nearly 70 trillion possible combinationsofchromosomesbetweenthetwoparents.Toputthisanother way, because of random fertilization, two parents would have to produce more than 70 trillion children before they produced two with the same set of chromosomes! That's a lot of potential diversity among offspring.

To sum it up, there are three ways that meiosis generates genetic diversity; crossing over between homologous chromosomes in the tetrad, random assortment of the homologous chromosomes prior to the first round of cell division, and random fertilization of the gametes. These three sources of variation create an incredible amount of diversity by shuffling genes around. It's much like playing poker with a 52-card deck. With each round, the same cards are shuffled and dealt, but the players are dealt new hands each new game. Different combinations of cards are more likely to win than others. It's the same with different combinations of alleles, some are more likely to be successful.

Why Have Two Sexes?

If sexual reproduction perplexed scientists, having two sexes was even more paradoxical. Why would a system of reproduction evolve two distinct sexes where you immediately cut in half the number of potential mates? Why not just have everyone be the same sex, that way you would have an easier time finding a potential mate, reducing the probability of being outcompeted by asexually reproducing organisms. Despite the potential shortcomings, sexual reproduction using two different sexes must impart some advantage because it is quite common among plants and animals.

To answer the origin of two sexes, we can once again turn to the mitochondria who seem to be responsible for many of the uniquely eukaryotic traits. If you are living inside your host cell, and it becomes damaged, it could be beneficial to bring in a second copy of a genome. With two copies of the genome, you are likely to have at least one copy of a gene that works. If one copy of a gene becomes damaged, it becomes easier for DNA repair mechanisms to find and fix any damaged DNA.

But, how do you bring in a second copy of a genome? The easiest way is to fuse two cells together, and this is what happens with gametes during sexual reproduction. Meiosis evolved to produce haploid gametes, so that the number of chromosomes does not continually double with each generation. However, the two gametes are not equal in their size, or contribution to the next generation. In most eukaryotes, it is generally easy to determine the female sex from the males by simply looking at the gametes (except for fungi, they do things a little differently, but I won't get into that here). In most cases, females produce fewer large eggs, whereas males produce many tiny sperm. Not only are the eggs larger, they contain all the organelles and mitochondria, whereas the sperm are usually tiny stripped-down cells carrying a copy of the father's genome. Basically, male gametes are nothing more than a source of genes.

Why do we inherit our organelles and mitochondria only from our moms? If you're a mitochondria living and reproducing asexually inside a cell, then all the other mitochondria are 100% related to you. There may be hundreds, or event thousands of genetically identical mitochondria in a single cell. If another cell were two fuse with your host bringing in new mitochondria, you would not be related to each other and then you would be competing against each other for resources. The evidence suggests that the evolution of two sexes was likely driven by mitochondria not wanting to compete with unrelated mitochondria from another cell. To point out the obvious, this system works because two sexes is practically ubiquitous across all eukaryotes.

Sexual Selection

Sexual selection has led to acrobatic dances, bizarre displays of brightly colored feathers, and other exaggerated features that require energy to maintain and practically scream to predators, hey come and eat me! Sexual selection occurs when one member of the other sex chooses a mate based on features that may not impart a direct survival advantage. Examples of sexual selection abound in the animal kingdom, ranging from the subtle, such as a small orange spot on a fish fin to the practically absurd of peacock feathers, all in the name of attracting mates.

Despite the costs of fancy ornamentation or dancing in the open, sexually selected traits have repeatedly evolved in numerous groups of animals, often with some spectacular examples. One theory to explain why sexually selected traits exists is based on the good genes hypothesis. It goes something like this; brightly colored feathers indicate that you have good genes and are otherwise healthy. It also means that you are well-fed, mostly disease and parasite free, and able to avoid predators even though you stand out. Maintaining colorful ornamentation or conducting fancy dances lets members of the opposite sex know that you are a good potential mate. In recent years, scientists have studied both female and male choice. Although female choice has been the most popularly studied, one species we are all familiar with, has undergone obvious male choice as we shall see.

Female choice is a type of sexual selection when females choose males based on certain traits that can seem almost completely arbitrary. In some cases, it almost seems totally random why females chose a certain trait over another, it appears as random as fashion trends in our society. For example, in the 1980s neon colors were all the rage, and in the 1990s everyone was wearing faded blue jeans and plaid shirts. The end result is that female choice has led to some unusual behaviors and colorful ornamentation that serves no purpose in survival.

As a birder, I'm always drawn to the ornate feathers of birds, they provide a wonderful example of female choice. Much of the bright coloration in birds is the result of females choosing males with the most attractive colors. In most birds with brightly colored males, the females remain drab compared to their male counterparts, mostly because it's in their best interest to remain somewhat camouflaged while incubating eggs on a nest.

A great example of the power of female choice to create over the top ornamentation occurs in the Indian Peafowl with their bright colors and spectacular tail feathers that are longer than the bird itself. All their feathers are beautifully colored in many shades of blue and green, they even have small bright blue ornamental feathers on their head. These brightly colored feathers don't necessarily help males with their day to day survival, but they do help attract the ladies! Think about it this way; if you're a male peafowl and can avoid predators, find food, and maintain those beautiful feathers, then you're a stud and the females will choose you because of your good genes. I should point out that the very large tail of peafowl may actually reduce predation by literally scaring away potential predators.

Not all females are attracted to brightly colored males, in many songbirds, females often select males based on quality of their songs. The Northern Mockingbird provides a great example where the males and females look basically identical. However, the males are quite vocal as they sing intricate songs known to include the various sounds in their environment. Male birds begin to learn their songs while still in the nest and continually add new sounds to and improve their songs throughout their lives. One particular mockingbird in north Florida learned how to mimic the author's alarm clock, which it liked to repeat every morning at the crack of dawn and sometimes in the middle of the night.

For some female birds, singing isn't enough, they require intricate songs and a display of beautiful feathers. Perhaps the greatest bird mimic is the Superb Lyrebird found in Australia. It has the ability to faithfully reproduce practically any sound in its environment including other bird calls, clicks from cameras, or even nearby chainsaws. While singing, it provides a display of its feathers for admiring females. On lakes in western North America, female Western Grebes do not rely on fancy feathers or complex songs to find a mate. Their courtship relies on the ability of the male to perform intricate dances in unison while running on water.

Yet, for some female birds, they want it all: beautiful feathers, singing, and dancing. In prairie chickens, found in the grasslands of the Midwest, the males gather together in large groups called leks. When lekking, the males perform intricate dances, sing, and display their feathers, all to attract a mate. What's remarkable is that these males perform in the open starting at daybreak. I imagine it is tough being a male prairie chicken having to maintain so many different sexually selected traits to attract a female, all while avoiding predators.

Female choice is also found in other groups of animals. In darters, small freshwater fish found in clear streams, the males possess colored fins with varying degrees of reds, blues, and greens that females must find appealing. In a group of new world lizards known as Anoles, the male has a highly colored throat patch, called a dewlap that he displays to the females and other males. Some male spiders are much smaller than the females who build webs. The males often live in small corners of the web and when the time is right, they must approach the female with great care. Each step must be perfectly choreographed, or she will instantly pounce and eat him.

While examples of female choice are easy to find in nature, there are fewer examples of male choice, but there are some obvious ones close to home. It's well documented that animals spend a lot of time and energy attracting potential mates, you don't have to go the wilds of the tropics to witness this phenomena, just go to your local night club and you will witness ornamentation, posturing, and sexually selected traits on display. Humans, like other animals have sexually selected traits and one of the best examples of male choice are female breast. The ancestors to modern male humans most likely chose enlarged breast because it was as sign of good genes and fertility. However, adipose tissue surrounding the mammary glands is not required for milk production. In addition to enlarged breast, there is evidence that men also selected for fuller lips and a certain hip to waist ratio.

Indeed, human males have evolved to be visually attracted to certain traits of female bodies and the science fully supports this observation. When men look at attractive women, or are shown pictures of breasts, regions in their brain associated with rewards are strongly activated. Additional studies have also revealed that men score lower on IQ tests when pictures of attractive women are present because they are distracted and unable to concentrate.

Malaria
A Simple Organism with a Complex Life

Most of us are familiar with sexual reproduction, a male and female get together, they produce gametes which then fuse to form a new organism. However, not every organism has such a simple life cycle. In fact, many parasitic organisms have complex life cycles with one or more stages reproducing asexually and another stage reproducing sexually. To complicate the life cycle even more, a parasite may depend on more than one host to complete its life cycle, while it uses different reproductive strategies in each host. In most cases parasites may reproduce asexually in their secondary host and reproduce sexually in their definitive host.

I'll use malaria as an example of a parasite with a complex life cycle because it has been and continues to be one of the greatest health threats in the world. Luckily for those of us living in North America, the winters are cold enough to prevent the spread of malaria and we have largely eradicated this disease in the south by intensive insecticide spraying. Malaria is caused by single-celled eukaryote in the genus *Plasmodium* and spread by mosquitoes in the genus *Anopheles*. When infected, the symptoms include fever, headaches, and vomiting, which can lead to coma and death. It takes about 10-15 days for the first symptoms to appear and recurrences can occur months later.

Despite the suffering caused by malaria, the life cycle of *Plasmodium* is quite interesting and I present it here just to show that some organisms have evolved very unique ways to reproduce and that humans are not the center of the animal world. I encourage you to read through this a couple of times, the terminology may be new, but focus on the overall picture. *Plasmodium falciparum* is one species that causes particularly severe cases of malaria. Similar to other eukaryotic organisms it reproduces sexually (meiosis) and asexually (mitosis). However, *Plasmodium* has several different forms in its life cycle depending on where they are found

I'll start the life cycle of malaria with the infection of a person. It begins when a female mosquito bites someone allowing *Plasmodium* to enter the blood stream. The parasite goes to the liver where it reproduces asexually by mitosis inside the liver. Over a short time, it will produce thousands of clones that eventually exit the liver and enter the circulatory system. Once there, they begin to infect red blood cells where they undergo a second round of asexual reproduction to increase their numbers and infect more red blood cells. This is important because the more red blood cells that become infected with *Plasmodium*, the more likely they will get into another mosquito. Humans are the secondary host for *Plasmodium*, we only harbor the parasite for a short time where it reproduces asexually to improve its chances of reaching the next stage of its life cycle. It may be hard to believe this, but we are basically disposable to the parasite. We are just there to ensure that the parasite gets into a female mosquito.

Once the parasites enter into another female mosquito they reproduce sexually by fusing to form a diploid zygote, or a fertilized egg. Once the zygote is formed, it will then divide by meiosis to form a haploid version of the parasite that will migrate to the saliva of the mosquito to be transmitted to the secondary host. And the cycle continues.

The female mosquito is the definitive host for malaria, which is defined as the host where sexual reproduction takes place. To the *Plasmodium* parasite, the survival of the mosquito is more important than the survival of humans, the secondary host. As you can see, humans are less important to *Plasmodium* than the mosquito, that's why humans infected with the parasite often die because it's in the interest of the mosquito to cause a higher rate of infection to ensure passage to the final host. Mosquitos typically don't suffer from the *Plasmodium* parasites.

The Birds and the Bees
It is a Wild World

I end this chapter with some of my favorite examples to illustrate just how varied the animal kingdom is with regards to sex determination and mating strategies, sometimes even forsaking sexual reproduction. I named this section after an old saying about growing up, "it's time to learn about the birds and the bees." I've never understood this saying, especially as a biologist, especially because the way birds and bees reproduce is nothing like humans, except that some female birds prefer good looking males capable of singing and dancing. Let's begin with the bees because their reproduction is about as different from humans as you can imagine.

Male honey bees (*Apis mellifera*) do not have fathers, but they do have grandfathers. The same goes for ants, a close relative of bees, in fact you can think of ants as a type of wingless bee. As you may have noticed, many bees and ants live in large colonies with a single queen laying eggs. Most of the individuals are female workers that do not reproduce. The males are drones and have one function, mate with the queen. Here's how it works; female bees and ants are diploid, meaning they are formed by sexual reproduction, they have a father and a mother. Males are haploid, they only have one set of chromosomes from their moms, the queen. When the queen lays a fertilized egg, it develops into a diploid female worker; when the queen lays an unfertilized egg, it develops into a haploid male offspring. In this system, known as haplodiploidy, the male drones are 100% related to the queens and the female workers are 75% related to each other compared to human siblings that are on average 50% related to each other. Evidence suggest that the close relatedness of the colony allowed them to evolve complex social structures, which is rare in the animal kingdom.

"Stop parthenogenesis, have sex", pronounced a t-shirt one of my undergraduate friends used to wear. Parthenogenesis, the ability to reproduce asexually, has independently evolved in many animal groups except mammals. The New Mexico Whiptail (*Aspidoscelis neomexicana*) is a small, ground-dwelling striped lizard where every individual is female. It turns out that the females have forsaken sex, gotten rid of males, and reproduce asexually by producing genetically identical female offspring. The eggs remain unfertilized, yet grow into adults. But, there is a catch, these female lizards have to simulate copulation in order to induce egg formation. Lizards aren't the only animals having evolved parthenogenesis. Aphids, small insects that feed on plants, also reproduce mostly by parthenogenesis. By the time the females hatch from an egg, they are already pregnant with the next generation!

The evolution of parthenogenesis goes to show that evolution by natural selection is not evolving towards a goal. Presumably, parthenogenesis evolves because there are immediate benefits associated with not having to find suitable mates, allowing for rapid reproduction. However, because many parthenogenetic species are relatively young, it supports the theory that these species are likely doomed to an early extinction without sexual reproduction generating genetic variability.

In humans, gender is set very early in development, but this is not so for many species of fish, especially in the tropics. In the case of grouper, a popular gamefish, they are all born female. As they grow larger, some will mature and become males. Groupers may live in small groups where the largest female will change sex to become a male. Approximately 75% of fish that switch sex follow this pattern.

The clownfish, popularized by Disney's movie *Finding Nemo*, also goes through a sex change during life. Small groups of male and female clownfish inhabit sea anemones and are immune to its stinging tentacles. Males are typically smaller than the female. If the female is removed or dies, the male will grow larger and become a female. Imagine Disney having to explain to young children how Nemo's Dad became his Mom!

The Bluehead Wrasse (*Thalassoma bifasciatum*) complicates the picture even more. They can begin life as a male or a female, but it is the females that when they grow larger change sex. This is not the end of the story for this fish, they also go through several phases of life known as initial phase and terminal phase. Larger females will switch to males and enter the larger terminal phase and are capable of holding territories for spawning. The smaller initial phase males possess larger testes than the terminal phase males so they are able to compete with the larger males even though they cannot guard a larger territory. In this case, the smaller males sneak in and quickly deposit their sperm when the larger males are spawning. These smaller males are affectionately known as sneaker males, proving that you don't have to be the largest or strongest male to be successful at mating.

As we've seen in fish, some animals can switch sex at some point in life based on environmental ques. In humans, sex is determined by the two sex chromosomes, the X and much smaller Y chromosomes. XX means you are female and XY means you are a male. Other animals including some birds, fish, and reptiles have a Z and W chromosome where males are ZZ and females are ZW, the opposite situation of mammals. We learned about haplodiploidy in ants, bees, and wasp where males develop from unfertilized eggs and are haploid, whereas females develop from fertilized eggs and are diploid. In crocodiles and alligators, sex is not determined by sex chromosomes at all; rather their sex is determined by the temperature at which the eggs incubate. Eggs in warmer parts of the nest develop into males, whereas eggs in cooler regions of the nest develop into females.

Now for something even more complicated. A group of bacteria parasites in the genus *Wolbachia* are found in many insects. Although, in some cases the relationship may be more mutualistic with both organisms benefiting from each other's presence. For example, some species of insects cannot even reproduce without the bacteria being present. Most types of *Wolbachia* are transmitted to the next host through the eggs of the female. To increase their transmission rate to their next host, the bacteria has been known to kill males while still larvae or turn males into females, and induce parthenogenesis in females, forsaking sex all together.

There are many other stories here; in fact, careers have been built on studying how life reproduces from understanding the genetic bases of sex to the evolution of different behavior and mating strategies across the species. But what's amazing, is that despite the enormous variety of behaviors and sex determination, one thing remains, sexual reproduction with two sexes evolved in our distant eukaryotic ancestor and has not only been maintained, but has diversified as well.

Chapter 7

The Unseen World
of
Single-celled Organisms

In a single spoonful garden soil, there are more
bacteria than there have been people that have ever
lived on this planet.

Don't think for a second prokaryotes are primitive; they
have been evolving for perhaps 3.7 billion years.

Introduction

The first fully independent living organisms on Earth were structurally simple prokaryotic cells, for life started out simple by necessity. Indeed, the name prokaryote means "before the nucleus", an accurate description as prokaryotes may have existed for some 1.5-2 billion years prior to the arrival of the larger and structurally more complex eukaryotes. Even today, prokaryotes still dominate the Earth in sheer numbers and perhaps biomass too.

When it comes to diversity, prokaryotes are often passed over due to their seemingly simple structure, which makes it difficult to tell different species apart by simply examining them under a microscope. However, we are in a golden age of genetic discovery brought about by rapid DNA sequencing techniques. For the first time, we can search for new bacteria by extracting tiny fragments of DNA from bacteria living on a grain of sand, or on small sample of human tissue. The findings have been nothing short of remarkable as we have discovered that prokaryotes are everywhere and much more diverse than we ever thought. While appearing structurally simple, prokaryotic diversity lies in their metabolism where they are so diverse they can extract energy from rocks, thermal vents, and sunlight.

It may seem odd, but eukaryotes can thank their origins to the coming together of two prokaryotes almost 2 billion years ago. What ensued was the largest restructuring of cell complexity in nearly 2 billion years as eukaryotic cells evolved internal membranes to compartmentalize various functions of the cell. This single event paved the way for multicellular life including plants, fungi, and animals.

Prokaryotes Paved the Way for Life

Life started out simply, because it had to. For nearly two billion years, life remained structurally simple, but the impact of single-celled prokaryotes to the Earth cannot be understated. To survive, life needs energy, therefore it must interact with its environment to acquire energy necessary for survival. Over time, microbial evolution led to new metabolic pathways allowing life to extract energy from the environment in novel ways.

Based on the fact that basically every cell uses proton gradients to make ATP, the energy currency of life, it is thought that the first cells most likely harnessed the potential energy generated by natural proton gradients from alkaline vents. Once these early cells left their places of origins, they probably acquired energy from organic molecules in their environment. However, the most abundant and practically endless source of energy in the world is sunlight. It didn't take long for life to evolve that ability to tap into this energy source. As a result, the ability to carry out photosynthesis evolved early in the history of life.

Photosynthesis is widely known for its crucial role in transforming the energy from sunlight into chemical energy stored in carbohydrates, thus making it available to life. But it also makes carbon available to life by fixing carbon dioxide into organic molecules. However, the impact of photosynthesis does not stop here. The byproduct of photosynthesis is free-oxygen (O_2), over billions of years the free-oxygen released from photosynthesis transformed the atmosphere into an oxygen rich mixture of gases that energized life.

Once atmospheric oxygen levels began to reach critical levels, the oxygen became toxic to many bacteria. But some prokaryotes evolved the ability to deal with the toxic effects of oxygen by harnessing its electron grabbing tendencies. The first aerobically respiring bacteria did something quite remarkable, they harnessed oxygen's electron grabbing tendencies to make more ATP. This one small act not only neutralized the toxic effects of oxygen by forming water, but it also generated much more ATP for every organic molecule the cell acquired.

The result was that oxygen energized these early cells, they were much more efficient at extracting energy form their environment and therefore had more energy to work with. It was the evolution of aerobic respiration in bacteria that paved the way for eukaryotic cells and eventually multicellular life. In fact, it was the merger of two prokaryotic cells, an aerobically respiring bacterium and an archaeon that led to origins of eukaryotic cells.

The presence of oxygen in the atmosphere is also important because high up in the atmosphere (10km to 50km) it reacts with incoming solar radiation to form a protective ozone layer. Ozone is comprised of three oxygen atoms (O_3) and filters out harmful UV rays from the sun. If it weren't for the ozone layer, there wouldn't be any life on the surface of the planet because it would be sterilized by all the UV rays. The Earth's oceans would also risk disappearing because ultraviolet light breaks water into oxygen and hydrogen gas. Our planet isn't quite large enough to hold on to hydrogen gas, so it escapes into space. Without an ozone layer, the Earth's oceans could slowly disappear, making the planet a very dry place, and not very habitable for life. Just look at the dry surface of Mars. It once had water on its surface, but it is now gone. Based on what we see on the Earth; some scientists think that life never had a chance to stabilize the Martian atmosphere. We can thank photosynthetic microbes for stabilizing our atmosphere, protecting our oceans, and making it possible for multicellular life.

Prokaryotes are Difficult to Classify

Human nature has a need to classify objects into discrete identities. Think about it this way, the borders of our country, states, congressional districts, and counties are clearly marked on maps. You live in New Mexico or Colorado based on which side of a line you reside. North of the state line, you live in Colorado, south of the state line, you live in New Mexico. There really isn't a middle ground. But, to migrating birds, or elk living in the San Juan Mountains, the habitat is continuous and our political boundaries are meaningless. When it comes to animals, we have attempted to classify them into a nested hierarchy from the most inclusive of kingdom (similar to a country) to the least inclusive of species (a county). In fact, many biologists will argue that a species is a unique and discrete unit, different from all other species.

However, nature often defies our attempt to neatly classify the variety of life into specific categories. Nowhere is the problem more prevalent than in the prokaryotes because they really don't fit into a nested hierarchy of kingdom, phylum, class, order, family, genus, and species. One way to classify organisms or taxa is based on grouping them by evolutionary relationships of taxa, which we can illustrate as a phylogenetic tree. We construct these relationships based on morphological and genetic similarities. The more two organisms have in common, the more recent their common ancestor and the more closely related they are to each other. Once again, the prokaryotes make this task incredibly difficult due to their large pangenomes and ability to swap parts of their genome in horizontal gene transfers.

Prokaryotes reproduce asexually producing genetically identical daughter cells, with the exception of random mutations that introduces a little variability. But, as you learned in the last chapter, horizontal gene transfer is rampant among unrelated prokaryotes and apparently regardless of how closely related they are to each other. While a typical *E. coli* may possess 5,000 genes, the entire pangenome may include 16,000 or more genes making it difficult to neatly classify them into a discrete species. In animals it's a bit easier, but could you imagine routinely swapping genes with organisms as different as a tree or a mushroom? That is exactly what prokaryotes do all the time. There is more genetic variation in *E. coli* in your gut than there is between you and a lizard, and we haven't shared a common ancestor with a lizard in over 315 million years. As you can see, it can be quite difficult to classify bacteria.

Carl Woese in the 1970s was among the first to realize the immense diversity of prokaryotes. At the time, most scientists divided life into five kingdoms (bacteria, protists, fungi, plants, and animals) with four of the kingdoms comprised of organisms with eukaryotic cells. Woese used advanced molecular techniques for the time to study the similarities of ribosomes between the five kingdoms. Recall that ribosomes are complex molecular structures used to make proteins in all cells. Their general structure is highly conserved because any change in its structure would almost certainly be harmful. But, over time, some small changes do occur and are kept and passed down through the ages. Because ribosomes and the genes that code the information to make them are highly conserved, they can be used to reconstruct evolutionary relationships of very different organisms that diverged millions to even billions of years ago.

The results of Woese's study showed that a relatively small unknown group of prokaryotes living in extreme environments were vastly different from most other prokaryotes. He also discovered that there were far fewer molecular differences between the four kingdoms of eukaryotes, the protists, plants, fungi, and animals than there were among the prokaryotes. The results were ground-breaking, leading him to propose a new classification system consisting of three domains comprised of bacteria, archaea, and eukarya. Four kingdoms of eukaryotes were combined into one domain, while the deceptively simple bacteria were divided into two domains, bacteria and archaea. The new classification was based primarily on differences in the ribosomes. Despite the vast outward differences between plants and animals, at the molecular level, all eukaryotes are very similar.

Soon after Carl Woese's study, there was renewed interest in studying prokaryotes, and it became easier with new technologies. What we have discovered regarding the differences between bacteria and archaea have supported Woese's study. To start, major differences exist between the structure of their cellular membranes. They are both made of phospholipids, but they use different building blocks requiring unique metabolic pathways to create their phospholipids. Using different building blocks to construct cellular membranes may not seem like much, but scientists studying the origins of life have speculated that the differences are a strong indication that life acquired their cellular membranes twice and independently; once for the bacteria and once for the archaea. This means that bacteria and archaea began to diverge at the origin of life itself.

It remains unclear just how many different types of prokaryotes may exist in the world. To further appreciate prokaryote diversity, check out these facts. Genetic studies analyzing the bacterial diversity in a handful of soil has estimated that there could be upwards of 8 million species of bacteria. Studies of leaves from one species of tree revealed several hundred species of bacteria, which were different compared to the bacteria on the leaves of another tree. We already know that a single human may have a thousand species of bacteria living in their gut, and there may be thousands of different species of bacteria capable of living in our gut, and we are just one type of animal. To put this diversity into perspective there are only 5,400 species of mammals, about 10,000 species of birds, 300,000 species of plants, and about 2 million species of animals. A single handful of garden soil may house the same amount of prokaryotic diversity as all the plants and animals in the entire world!

Prokaryotes are Metabolically Diverse

Don't call prokaryotes primitive organisms just because they lack a cell nucleus and remain otherwise structurally simple. What prokaryotes lack in structural complexity compared to eukaryotes, they make up for it in metabolism. Prokaryotes were the first organisms to evolve photosynthesis nearly 3.5 billion years, forever transforming the surface of the Earth by releasing free-oxygen to the atmosphere. In turn, this paved the way for aerobic respiration, the ability to use oxygen as a final electron acceptor vastly improving the efficiency of ATP production. Additionally, the origin of eukaryotes occurred when two very different prokaryotes merged together with one cell living inside the other in a mutually beneficial relationship. This unlikely merger ultimately paved the way for active multicellular animals.

All life uses metabolic processes to create order by taking in energy and materials from the environment. Abiogenesis can trace its roots back to chemical evolution beginning with the simple reaction of adding hydrogen to carbon dioxide, or fixing carbon. Broadly speaking, organisms can be heterotrophs (hetero=other, troph=feeding) or autotrophs (auto=self, troph=feeding). Heterotrophs can only acquire nutrients and energy from organic molecules, typically made by autotrophs. Autotrophs are self-feeders capable of extracting energy from the environment and using it to make all the organic molecules they need by fixing carbon dioxide. Organic molecules are based on the element carbon because of its ability to form four stable chemical bonds allowing it form millions of different organic molecules.

There are only a few metabolic process that add hydrogen to carbon dioxide and autotrophic prokaryotes evolved them all. Recall that metabolism is the sum of chemical reactions taking place in cells that bring them to life. Some reactions break down molecules into smaller ones, often accompanied with a release of energy that cells can harness to do work. In turn, the smaller molecules can be repurposed as the building blocks for larger more complex molecules. Next, I explore some of the microbial diversity in our world starting with chemoautotrophs, a type of organism that is capable of extracting energy from the environment to fix carbon dioxide.

Chemoautotrophs - Methanogens: You may have never heard of an Archaean known as a methanogen, and few people have. Yet, they may be one of the most important organisms to our evolution you never heard of. Based on current research, they may have been the cell that an aerobically respiring bacteria took residence in over 2 billion years ago, jump starting the evolution of eukaryotes! Methanogens are autotrophs that react hydrogen gas (H_2) with carbon dioxide (CO_2) to form methane (CH_4), an organic molecule found in natural gas. This is an example of carbon fixation when inorganic carbon is converted to organic carbon by the addition of hydrogen atoms.

They say there is no such thing as a free lunch, but that's not entirely true all the time, at least for methanogens. Hydrogen gas has a lot of potential energy, the bonds are easy to break; it practically wants to react with other molecules, especially molecules with oxygen. Recall that electrons are used to make chemical bonds. With a little help from a metal catalyst, hydrogen gas can react with carbon dioxide to form methane and it releases a little bit of energy. It's like getting paid to eat lunch! What this means is that this reaction for fixing carbon dioxide is actually energetically favorable.

If you've ever walked into a swamp, or heard of swamp gas, it's methane produced by these tiny microorganisms. In recent years, archaea have been discovered to be part of our gut microbiome. Some people, have methanogens in their gut producing methane, similar to swamp gas. The implication is that their flatulence is flammable.

Chemoautotroph - Iron Bacteria: Have you begun to notice the importance of energy for living cells? Electrons can serve as source of energy, hence the connection between electricity and electrons. Even the electricity in your home is the result of electrons moving in the wires from the positive end to the negative end. Some bacteria can literally extract energy from rocks by snagging their electrons. Specifically, a form of reduced iron, known as Iron 2 (Fe^{++}), donates an electron carrying energy that some cells can use. These cells uses the energy in the electrons to fix carbon dioxide into organic molecules. In chemistry, reduced means you have electrons, whereas oxidized means you've lost electrons. If you've watched iron rust, it's slowly being oxidized, losing its electrons to oxygen. Prior to the oxygen revolution brought about by the photosynthetic action of cyanobacteria, much of the iron was reduced and able to donate electrons to iron bacteria. Iron bacteria rely on the reduced form of iron found inside of rocks, or rocks that haven't been exposed to the air for too long. In addition to iron, bacteria can also grab electrons from other minerals as well.

If you've ever been in the mountains or near sites with mining, you may have noticed bright orange streams from acid mine drainage. Iron bacteria are responsible for acid mine drainage from mines, a major problem for stream quality. As mines age, they slowly fill with water. Bacteria and archaea begin to grow and oxidize (remove their electrons) the minerals, slowly lowering the pH and oxygen levels of the water. These bacteria actually favor the low pH conditions and are a type of extremophile known as an acidophile. The bright orange color is the result of the oxidized minerals, mainly iron minerals.

Photoautotrophs - Cyanobacteria: Photosynthesis uses the energy in sunlight to fix carbon dioxide into carbohydrates. Recall that every cell on this planet uses a proton gradient to generate the majority of its ATP. Through evolution, cyanobacteria were able to repurpose the proteins involved in generating proton gradients into photosystems capable of transforming the energy in sunlight to chemical energy. The origin of plants began about one billion years ago when a eukaryotic cell engulfed a cyanobacteria. Rather than digesting it, the cyanobacteria became an endosymbiont along with the mitochondria forming the first eukaryotic algae. One lineage of algae went on to evolve into plants, which dominate terrestrial ecosystems. To emphasize the point here, plants did not evolve photosynthesis, bacteria did.

Reverse Krebs cycle: For most biology majors, remembering the steps of the Krebs cycle is a rite of passage (although, the details soon forgotten after an exam): a series of eight chemical reactions where bonds are broken and new ones are formed, thus creating new molecules at each step. Recall that the Krebs cycle takes place in the mitochondria and its purpose is to break down organic molecules by stripping them of their electrons, forming carbon dioxide as a waste product. The high energy electrons are used to make ATP for the cell.

Having taken any intro-biology course, you would have been led to believe that the Krebs cycle exists as part of cellular respiration primarily to generate ATP. However, if you were to look at any metabolic chart, you would notice a circle representing the Krebs cycle in the middle of the chart. The molecular intermediates generated from the Krebs cycle can be siphoned off to be used as the building blocks of lipids, carbohydrates, proteins, nucleic acids, and various other molecules required by the cell. The Krebs cycle is ubiquitous among all living organisms betraying its ancient ancestry to the first cells. Its central importance to cellular metabolism indicates it may have evolved through chemical evolution before the first cells had membranes.

But, what you may not know is that if you add a source of energy to the Krebs cycle and hydrogen gas (H_2), it can add hydrogen to carbon dioxide! The Krebs cycle is a metabolic pathway that can fix carbon by adding hydrogen to carbon dioxide, meaning it can run backwards. Rather than being catabolic, only capable of breaking down organic molecules, with an input of energy, it can run backwards allowing the organism to become autotrophic. However, it's not well known how often this happens.

Nitrogen fixers: The Earth's atmosphere is about 78% nitrogen (N_2), which is also one of the four most abundant elements found in living organisms. Even though it is quite abundant in the atmosphere, nitrogen gas is inert and not very reactive making it practically unavailable to most living organisms. For life to use nitrogen, it must be "fixed" into molecules available to organisms, much like carbon must be fixed into organic molecules using energy. Nitrogen fixation also requires an oxygen free environment and energy to fix nitrogen into ammonia (NH_3) or nitrates (NO_3). No eukaryote (plants or animals) can fix nitrogen, but bacteria can. However, plants in the pea family possess root nodules that house a symbiotic bacteria called *Rhizobia* that can fix nitrogen into usable forms. The roots provide an oxygen free environment and energy while the *Rhizobia* bacteria provide usable nitrogen to the plant.

Prokaryotes Live in Extreme Environments

Prokaryotes are everywhere, including some of the most bizarre and extreme environments you could imagine. They are found thousands of feet below the surface, living in aquifers and extracting energy from rocks. They have been found living in water hot enough to boil, or in freshwater lakes thousands of feet below glaciers, or living in rocks that are perpetually frozen in the Antarctic. As you just read, some iron bacteria can live in very acidic conditions in mines as they slowly oxidize rocks for their electrons. Below, I describe a few examples of prokaryotes living in extreme environments.

Extreme cold: In the 1990s, a team of researches braving the cold and harsh environment of the dry valleys in the Antarctic discovered bacteria eking out a living inside the rocks. We usually think of the Antarctic continent as being totally covered in ice; however, there are a few places where precipitation is so low that the snow sublimates. Sublimation is when ice turns directly into vapor, and in the Antarctic it leads to the formation of dry valleys. The lack of precipitation combined with extreme cold make this one of the harshest environments on Earth. Yet bacteria known as endoliths, often referred to as "rock eaters", can survive here by stripping electrons from rocks as an energy source. A trade-off for such a harsh living is that these bacteria grow incredibly slow.

Scientific discoveries did not stop with finding bacteria in Antarctic rocks of the dry valleys. In 2014, scientists discovered other "rock eating" bacteria in a freshwater lake buried several thousand feet beneath Antarctic ice. Archaea and bacteria that live in extremely cold environments are known as psychrophiles. Astrobiologists study endolithic bacteria living in extreme environments to gain insight into how life may exist on Mars (or other planets and moons) where it is also cold and dry.

Extreme salinity: The Dead Sea and the Great Salt Lake in Utah have a couple things in common, but most notably, they lack fish, a direct result of extremely high salinity. Although, the name "Dead Sea" is not quite accurate, on the contrary, it is full of bacteria and archaea capable of surviving in these extreme environments. Known as halophiles, these microbes must spend extra energy maintaining a proper water balance to survive in highly saline environments. One method involves the production of protectants that help keep water inside the cell. Without these protectants, water would rapidly leave the cell through osmosis causing them to shrivel and die. Another method of keeping water inside halophiles involves the active pumping of potassium ions into the cell. Ironically, if you were to place a halophile in freshwater, it would immediately swell and burst from the water rushing into the cell.

Extreme heat: Yellowstone is known for its geysers, including Old Faithfull, along with numerous hot springs with water temperatures approaching the boiling point. Prior to the 1960s, it was generally assumed that life couldn't exists much beyond 135°F (55°C), then bacteria were found thriving in temperatures up to 160° even 175°F. One of these small unassuming thermophiles was named *Thermus aquaticus*, which literally means "hot water". An important adaptation for these bacteria is to have proteins capable of operating at hot temperatures without falling apart or denaturing. When a protein denatures, it changes shape and loses its ability to function. If you've ever cooked an egg, you have seen how heat alters proteins when the liquid egg white changes from clear liquid to a white solid. Those types of changes to proteins in a cell would render them useless leading to cell death.

The discovery of *Thermus aquaticus* turned out to be one of the most important discoveries for the field of genetics. Here's why. Every cell that replicates its DNA uses a large protein complex called DNA polymerase. In *Thermus aquaticus*, its DNA polymerase molecules operate normally under very hot conditions. This may not seem to be overly important; however, finding it led to a major advance in our ability to sequence DNA. In the early 1980s, a new technique called PCR was invented to amplify small amounts of DNA so that it could be easily sequenced. However, as part of the process, the DNA had to be heated and cooled repeatedly. Unfortunately, the heating process denatured the normal DNA polymerase used to amplify the DNA, rendering it useless. As a result, it was very difficult to isolate small segments of DNA to be sequenced. However, using the DNA polymerase from *Thermus aquaticus* solved the problem of using heat as part of the DNA amplification process and allowed scientist to easily amplify sequences of DNA.

The DNA polymerase from *Thermus aquaticus* is called *Taq* polymerase. Its patent was sold for $330 million dollars and is speculated to have generated over 2 billion dollars in royalties. Its discovery also led to the Nobel Prize in 1993. *Thermus aquaticus*, a small thermophilic bacteria living in a hot spring in Yellowstone National Park led to a revolution in genetic research. The discovery of *Taq* polymerase clearly shows the importance of finding, studying, and preserving biodiversity in the world because you never know what you may find!

Disease Causing Bacteria

The vast majority of bacteria are harmless, some are beneficial, and some may even be required for our existence. And, of course, there are bacteria capable of causing diseases. Unfortunately, pathogenic bacteria are responsible for some very bad diseases including tuberculosis, salmonella, tetanus, syphilis, staph infections, along with various types of food and waterborne illnesses. There are many bacterial diseases, but I'm only going to focus on two, tuberculosis and pertussis (whooping cough).

Tuberculosis is a highly contagious, airborne disease that has killed millions of people. In 2013, there were 9 million new cases of tuberculosis and approximately 1.3-1.5 million deaths due to the disease. Many of these new cases are in poor regions of the world lacking adequate medical care and access to vaccinations. Tuberculosis (TB) is caused by a bacteria called *Mycobacterium tuberculosis*, which infects the lungs and causes major damage. Widespread vaccinations have dramatically reduced TB outbreaks in developed countries including the US. However, where vaccinations are less common, or people refuse to get vaccinated, TB remains prevalent. Most TB infections can be successfully treated with proper antibiotics. However, drug resistant strains have evolved and threaten to spread around the world infecting people who have not been vaccinated.

Whooping cough, also known as pertussis, is another highly contagious airborne disease caused by the bacteria *Bordetella pertussis*. It was once widespread, killing between 10,000 and 20,000 people per year in the US, with higher mortality rates in infants and young children. Pertussis produces a toxin that stops the cilia in the cells that line your throat. As a result, people with whooping cough will go into major uncontrollable coughing fits. The first vaccine was developed in the 1930s and mortality declined to less than 100 by the 1980s.

With the accidental discovery of penicillin in 1926, survival rates against bacterial diseases and infections greatly improved. Combined with the widespread implementation vaccinations, some bacterial infections including diphtheria, tetanus, pertussis (Whooping Cough) and tuberculosis have continuously declined to the point where they have almost been eradicated forever. Vaccinations work because we have an acquired immune system, meaning it remembers every pathogen that has ever invaded your body, including ones that made you sick or weakened versions from vaccinations. By getting vaccinated, our immune system remembers the pathogen allowing it to quickly react to the particular pathogen (bacteria or virus), eradicating it before it makes us sick.

In an unfortunate trend, uninformed parents are opting out of getting their children vaccinated against these serious diseases. They erroneously believe that vaccinations may lead to autism or other mental impairment. However numerous scientific studies have resoundingly disproved the claim that vaccinations cause autism or any type of mental impairment. In fact, the *Journal of the American Medical Association* (JAMA), one of the most respected journals in the world called the connection between vaccinations and brain injury a "myth" and "nonsense". Just to be perfectly clear, VACCINATIONS DO NOT CAUSE AUTISM OR MENTAL IMPAIRMENT.

In 2015, an outbreak of whooping cough in California led state lawmakers to remove personal exemptions for the vaccinations for all children entering the public school system. Getting a vaccination only reduces your risk of contracting the disease and works best when everyone else around you is also vaccinated. If a large portion of the population is not vaccinated, they are at risk for the disease and they put others at risk as well. Some studies have even shown that the effectiveness of vaccinations does not last a lifetime, but wanes over time, sometimes lasting less than a decade.

Infectious diseases will almost always remain a problem, they continue to persist because not everyone gets vaccinated, and there has been a rise in antibiotic resistant diseases. As I discussed in Chapter 6, horizontal gene transfer allows bacteria to swap genes with each other including genes for antibiotic resistance. Because of the widespread use of antibiotics, especially in the meat industry, antibiotic resistant bacteria are on the rise and spreading. Scientists are having to routinely develop new drugs to not only fight emerging diseases, but also old ones that are evolving resistance to current antibiotics.

Beneficial Bacteria – Our Microbiome

Bacteria often get a bad rap, when most people think of bacteria, they often think of uncleanliness and diseases. The fear of bacteria is rampant, just walk down the soap aisle at the local super market and you will find labels proclaiming how effective their product is in killing bacteria. One of my favorite boast of a product's effectiveness was printed on a particular brand of dish soap, where it claimed to kill 99.9% of all bacteria. At first, this may sound very effective. But, if there were 10 million bacteria on your spoon and 0.1% lived, you would still have 10,000 bacteria, and they would now be more resistant to your soap! You just caused the bacteria in your kitchen to evolve to be resistant to your dish soap.

Fortunately, the vast majority of bacteria are harmless, you don't even know they are there. For example, in a single spoonful of dirt, there are more bacteria present than there are people living on the Earth. In reality, bacteria are incredlbly important for the functioning of our ecosystems and ourselves. Complex life and ecosystems could not exist without bacteria and archaea.

It turns out that animals may actually be dependent on symbiotic bacteria for their existence. Your body harbors a lot of bacteria; they potentially outnumber your cells 10 to 1. For every one cell that is you, there are perhaps 10 bacterial cells living in and on you. It has been estimated you have about 10 trillion cells that are you, and about 100 trillion prokaryotes, most residing in your gut. For years, we have known that *E. coli* was the most common bacteria living in our gut, along with a few other types of microbes. Modern techniques in sequencing DNA have revealed that our microbiome is much more diverse and potentially beneficial to us than we ever imagined a decade ago. The average person has over a thousand different types of bacteria residing in their gut alone, and it is in their diverse metabolic activities where we derive our benefits. We have also discovered that no two people have exactly the same gut flora. Studies are showing that environment, including diet and where someone lives, along with their genetics are very important for determining the types of microbes present.

Currently, the full extent of the diversity of gut microbes in humans is not well known. In fact, their effect on our bodies is just starting to come into light. We have known that bacteria help us break down our foods and they also make various vitamins and enzymes we may need if they are lacking in our diet. The sheer volume of some bacteria, including *E. coli*, may actually form a protective barrier in our gut preventing us from getting diseases. The influence of our gut microbiome doesn't stop here. In recent years, there has growing evidence that our gut microbes can influence our behavior. In studies using mice, there is indication that some microbes affect serotonin levels in the gut. Serotonin is a chemical used by the brain to relay signals to the body, this one molecule can affect mood, appetite, sleep, memory, and sexual desire. What this means, is that there could be a link between our moods, cravings, and health based on gut our microbes, which in turn are strongly influenced by our diet. Understanding how microbes influence us is only in its infancy and will continue to be a source of research for years to come.

In recent years, there has been a rise in the probiotic industry promoting various foods containing beneficial bacteria, or probiotics including those found in yogurt (although, they must have live active cultures). A serving of yogurt with live active cultures every day can have a very positive influence on your gut microbe by promoting beneficial bacteria including several species of *Lactobacillus*. Having good bacteria can help a wide range of conditions including chronic constipation, diarrhea, and inflammatory bowel disease.

Observational studies on yogurt indicated that people who regularly eat yogurt have lower rates of obesity. In an effort to understand how yogurt and probiotics effect obesity, scientists conducted a controlled experiment on mice where some received yogurt with live cultures and others did not. At first, the scientists noticed that the hair of the mice on a daily diet of yogurt intake was shiny, coinciding with increased follicle activity, a clear sign of improved overall health. However, the researches made another unsuspected finding, they noticed that the males were "projecting their testes outward" giving the mice swagger. Indeed, the males that were fed yogurt had teste sizes 5% heavier than the males not eating yogurt. Additionally, they sired more offspring. Yogurt also paid off for females because they had larger litters and more pups were successfully weaned. Follow up experiments will determine if there are similar effects in human fertility and to determine exactly how the probiotics were improving overall health.

Prokaryotes and the Search for ET

It's highly unlikely that we will find a technologically advanced extraterrestrial (ET) civilization similar to ours. However, the surprising ability of prokaryotes to survive in extreme environments on Earth has excited astronomers and given credence to a growing field of astrobiology. In 2015 NASA picked Europa, a moon of Jupiter with a liquid ocean under its icy surface as a destination for a new mission specifically to search for life. Later that same year, NASA confirmed liquid water on the surface of Mars, making it the second place in the solar system with liquid water on the surface. The finding is really exciting because it improves the chances that life could be present elsewhere in the solar system.

Future missions to Mars, or any place in the solar system, with the intent to search for life will be tricky because we could easily contaminate any planet or moon with our own microbes. NASA has plans to send a manned mission to Mars in the 2030s. They will have to be careful when searching for life, especially considering, each person would carry a 100 trillion bacteria! However, the payoff could be incredibly huge if life was found elsewhere in the solar system. It would provide biologists and chemists an independent origin of life to study, which could shed further light on abiogenesis, the origins of life from chemical evolution.

NASA is currently developing additional missions to search for exoplanets and additional signs of life. These include several large telescopes so sensitive will be able to detect the atmospheric composition of exoplanets. One sign of potential life would be to find an oxygen-rich atmosphere, which would be a great indicator of photosynthesis taking place, a uniquely living process!

Endosymbiosis Gave Rise to Eukaryotes

The origin of eukaryotes is unique compared to the origins of other groups of organisms. Based on Darwin's theory of evolution, we know life evolves over time and modern species are the result of descent with modification as populations accumulate many small changes over time. But, when it comes to the evolutionary origin of eukaryotes, it's a bit more complicated. Prokaryotes did not slowly gain complexity in their cells eventually evolving into the more complex eukaryotic cell. They have always been structurally simple and will remain that way.

Prokaryotes cannot evolve into a large complex cell only to haphazardly engulf some unsuspecting aerobically respiring bacteria. The main reason why prokaryotes will remain small comes down to energy limitations. Recall that prokaryotes make ATP by chemiosmosis across their cell membrane. If they were to grow larger, they would gain volume at a much larger rate than surface area, and it is surface area they need for generating ATP. Basically, as they become larger, ATP becomes increasingly scarce inside the cell, limiting the amount of energy available to them. Therefore, it took the rather unique event of two prokaryotes merging together to jumpstart the evolution of eukaryotes, a non-Darwinian theory championed by Lynn Margulis in the 1960s. This radical new theory was called endosymbiosis, where 'endo' means 'within' and symbiosis means 'living together'.

When Margulis first proposed the endosymbiotic theory for the origins of eukaryotes, it was met with harsh skepticism. Part of the problem was that the technology required to support her theory wasn't available yet. Over the next few decades, support for the endosymbiotic origins of eukaryotes slowly continued to grow. By the 1980s, a revolutionary new genetic technique called a PCR was invented, making it relatively simple to amplify small fragments of DNA so they could be easily sequenced. Sequencing DNA is determining its primary structure, or the sequence of the four nucleotides. Recall that genetic information is stored in the sequence of nucleotides, just like letters of the alphabet convey information based on their sequences. Once PCR was available, mitochondrial DNA was sequenced and it revealed its distinct prokaryotic origins. Today, it is well established that the ancestor to mitochondria were once free living bacteria capable of aerobic respiration.

With the theory of endosymbiosis firmly established, the next question centered on determining the identity of the host cell involved and how did the bacteria find its way into another cell? These are very important questions, but with some basic knowledge about eukaryotic and prokaryotic cells, we can begin to address the question of how it happened and who was involved. To begin, there are no prokaryotes with organelles or a nucleus. Whereas, modern day eukaryotes have a membrane-bound nucleus and a network of membrane-bound organelles used to compartmentalize the duties of the cell. Eukaryotes also have a complex cytoskeleton made of proteins that enable eukaryotes to move through their environment, maintain their shape, organize their organelles, divide, and engulf food or other cells from their environment.

We know that it was unlikely that a prokaryote engulfed another prokaryote. The process of engulfing large objects is called phagocytosis and is energy an intensive process requiring a complex cytoskeleton. So far, we have never encountered a prokaryote with a cytoskeleton capable of phagocytosis. And lastly, we also don't see prokaryotes that are intermediate in complexity compared to eukaryotes. What we do see are that some bacteria are capable of boring into other bacteria, and we have also found bacteria living inside of other bacteria. Based on this, we could predict that the ancestor to mitochondria may have bored their way into the host cell.

This brings us to the next question, who was the host cell? Based on 40 years of accumulated molecular data, we know that the original host cell may have been an archaea. First, archaea and eukarya possess ribosomes that are more similar to each other than either are to bacteria. Second, the DNA of archaea and eukaryotes are wrapped around small proteins called histones, whereas bacterial DNA is naked. With this information, a picture begins to emerge that eukaryotes formed when a bacteria bored its way into an archaea about 2 billion years ago. Additional evidence supports that the ancient archaea was potentially a methanogen, the same type of organism we met earlier in this chapter that makes methane from hydrogen gas and carbon dioxide.

That singular event led to the biggest reorganizations of cells in over 1.8 billion years! Endosymbiosis jump started the evolution of eukaryotes. Once this happened, Darwinian natural selection began to operate on those first eukaryotic cells, eventually giving rise to complex multicellular life forms including plants and animals. But first, let's explore a little eukaryotic diversity that doesn't include multicellular organisms such as plants and animals.

Eukaryotic Cells Are Structurally Complex

Eukaryotic cells, aided by the extra energy available to them, the benefit of multiple mitochondria producing lots of ATP, were able to make a giant evolutionary leap in their internal organization and size. Most eukaryotes are giant compared to the smaller prokaryotes, in some cases, thousands of prokaryotes could fit inside a single eukaryote. It takes energy to be so large, that's why mitochondria are present in almost every type of eukaryotic cell, including plants. Some eukaryotic cells may actually possess upwards of a thousand mitochondria in them. I should note that there are a few highly evolved parasitic eukaryotes residing inside animals that over evolutionary time, their mitochondria have become greatly degenerate because they don't require a lot of energy to exist in their host.

The cell nucleus is another distinguishing feature of eukaryotic cells. It is a double membrane similar to flattened sacs stitched together by proteins. DNA is located inside the nucleus and is associated with numerous proteins necessary to keep it organized. In fact, eukaryotes get their name, new kernel, from having a defined nucleus. Therefore, many people consider the nucleus to be the defining character of eukaryotes. However, from an evolutionary point of view, it is really the mitochondria that are the defining characteristic of eukaryotes because they were present first and were important for allowing the evolution of eukaryotes to occur.

In addition to the nucleus, eukaryotes compartmentalize the functions of their cells with a series of membrane bound structures called organelles. Organelles can be thought of as little organs similar to your organs that carry out different tasks in your body. Most organelles are involved in the manufacturing and break down of molecules required for the cell's functioning. With its many organelles working together, eukaryotic cells are very dynamic allowing them to quickly respond to changes in their environment.

Since their origins nearly 2 billion years ago, eukaryotes have diversified into many groups. As we will learn in the next three chapters, eukaryotes are the only domain of life that evolved into true multicellular organisms, including plants and animals, with genetically identical cells carrying out different functions for the good of the colony. Many lineages of eukaryotes have also remained as single-celled organisms, some are quite important for life on Earth, while others have evolved into pathogens and parasites with elaborate life cycles.

The Single-Celled Protists

Protist is a generic term for eukaryotic organisms that are not truly multicellular, most are single-celled organisms. However, a few can form large colonies including, slime molds, kelp, and red algae. These large colonies are not true multicellular organisms because they do not form distinct tissues that we see in plants, animals, or fungus. Despite this, there may be 200,000 species of protists in the world living in practically every environment, including our own guts alongside bacteria. Below, I discuss just a few different types of protists that I hope you find interesting.

Phytoplankton is Greek for "wandering plants", in some ways that's partly true because they float in water unattached to any hard surface, moving with the wind and tides. They are not quite plants, but they do have chloroplasts, the energy converting organelle used for photosynthesis, making them autotrophs. Phytoplankton are incredibly important to the entire world. First, they are the bases of most marine and freshwater ecosystems. If you've ever eaten fish, especially from the ocean, the carbon in that fish was probably fixed by phytoplankton. Second, it has been estimated that phytoplankton in the oceans are responsible for 50% of the world's oxygen production.

Although most phytoplankton are microscopic, too small to see with the naked eye, they are quite diverse in their shapes and sizes. Nowhere is this more evident in a group called the diatoms, perhaps the most abundant form of phytoplankton with an estimated 100,000 species. Diatoms show up everywhere; in rivers, lakes, marine ecosystems, and home aquariums. Diatoms are mostly found as single cells; however, some can form loose colonies. Their unique cell walls are made of silicates and resemble two separate shells (valves). The tiny silicate shells of diatoms come in a wide range of shapes and sizes, some of them are among the strongest features in the world. It's been estimated that the strength of some diatom valves, if enlarged, could support the weight of an elephant. Studying the intricate patterns of the diatom valves may provide engineers with new ways to build very strong structures. Diatoms may offer additional advances in biotechnology to produce technology at the nanoscale and to potentially help in the manufacture of solar panels.

Algae is a catch-all term for photosynthetic organisms including phytoplankton and cyanobacteria. If you've ever owned an aquarium or seen one, you have probably noticed algae forming a green film on the glass. There are thousands of types of algae living in practically every environment with water and sunlight, but they are limited to areas that are not too extreme in temperature or pH. In recent years, there has been an interest in farming algae as a source of biofuel to replace fossil fuels. Algae fuel would have several advantages over fossil fuels. One is that it would reduce the input of carbon dioxide into the atmosphere, lessening our impact on global climate change. Second, it would be renewable source of energy. To farm enough algae to supply the nation's oil needs would require about 1/7 of the area we use to currently grow corn. That's an area about half the size of Maine. The current obstacles to algae-based oil include engineering algae to cheaply produce oil and building the infrastructure to grow and harvest the algae made oil.

The single-celled *Euglena* has characteristics of both plants and animals. Well almost, it's not a multicellular organism, so in that way it is neither a plant nor an animal. They do have chloroplasts making them capable of carrying out photosynthesis. However, when there is not enough light to power photosynthesis, these single-celled organisms switch to becoming heterotrophs. Specifically, they begin to engulf other species of algae. As you can guess, it was difficult to classify *Euglena* because of their heterotrophic and autotrophic tendencies. Modern techniques allowing us to sequence DNA has shed light on their diversity and classification. There are at least 800 species of these protists found in most freshwater bodies worldwide, and in some cases in large enough numbers to turn the water green.

Sometimes the lines between multicellular organisms and single-celled organisms can become quite blurred, especially when the organism in question is the size of a mushroom and easily seen with the naked eye. I'm talking about slime molds, an informal name given to a type of protists capable of aggregating into large groups to reproduce. Most of the time, a slime mold is basically just a bunch of individual cells hanging out by themselves. Then some environmental que, such as a change in food availability, causes them to come together into a giant amorphous body. Once they come together, they are able to detect odors in the air and move towards it. Even though a slime mold is made of numerous individual cells, it can behave similar to a multicellular organism.

Animals almost certainly evolved from a single common ancestor some 600 million years ago. To determine what that ancestor was, we can study sponges, a unique type of organism that blurs the line between a colony of cells and a true multicellular organism. It turns out that sponges, a type of organism that most scientists classify as an animal. Sponges have specialized cells, but they lack true tissue layers, internal organs, symmetry, and specialized genes capable of forming a body plan. However, they do have collar cells whose function is to move water through the sponge.

Collar cells possess a flagella surrounded by a small collar, hence the name. There is a protist called a *Choanoflagellate* that also has a flagella surrounded by a small collar, appearing almost identical to the collar cells of the sponges. Most of the time, choanoflagellate are single-celled, but if they sense a predator in the area, they can form a small colony to avoid being eaten. Perhaps the origins of animals began when the ancestor to choanoflagellates began to form colonies to avoid predation.

Chapter 8

Plant and Fungal Diversity
and
The Greening of the Earth

The land was barren, devoid of plant and animal life for the first four billion years of Earth's history; then slowly plants colonized the land creating the first forests and the greening of the Earth.

Introduction

For the first 4 billion years of the Earth's history, the surface of the Earth was barren; no forests and no animals were to be found anywhere. Then, about 470 million years ago, the first plants and fungus began to colonize the land and they may have done so together. However, the story of plants actually begins about a billion years ago when the eukaryotic ancestor to plants acquired their photosynthetic ability by engulfing a cyanobacteria. Over time, these small photosynthetic bacteria eventually evolved into chloroplasts, the organelle where photosynthesis takes place in plant cells.

The story of terrestrial plants began as small mosses known as bryophytes evolved from algae about 470 million years ago. Since that time, plants have continuously evolved to thrive on land and grow taller by evolving lignified cell walls, leaves, roots, and a vascular system. It took the evolution all of these structures allowing plants to grow tall to form the first forest, forever transforming the surface of the planet. Another evolutionary milestone occurred in the early Cretaceous about 120 million years ago with the appearance of flowering plants. Today there are over 300,000 species of flowering plants alone, which would take many volumes of books to cover their amazing diversity.

Fungi most likely colonized the land at the same time as plants about 470 million years ago. They are multicellular and heterotrophic, a quality that has made them a vital component for the world's ecosystems as decomposers of dead plant and animal material. By producing a variety of enzymes, fungi can break down organic matter and even speed up the erosion of rocks, releasing their minerals and making them available to life. In addition to their roles as decomposers, some fungi are pathogens and have been implicated in amphibian declines, athlete's foot, crop destruction, and witch hunts. Not to mention that fungi are used to brew beer, bake bread, create antibiotics, while others can cause hallucinations and death! Although fungi have evolved into many species, their total diversity is not well known because many are microscopic.

Part I: Plants
The Origins of Plants

From the previous chapters, you know by now that it was the prokaryotes who evolved the ability to conduct photosynthesis. Recall from Chapter 3 that photosynthesis is a series of chemical reactions where the relatively inert gas carbon dioxide (CO_2) is "fixed" into organic molecules, and free oxygen (O_2) is released as a byproduct. To put it another way, photosynthetic organisms make their own food from carbon dioxide and water by using the energy in sunlight.

The origin of plants dates to about a billion years ago when an early eukaryote engulfed a cyanobacteria. Eventually the cyanobacteria would evolve into chloroplasts inside the cells and is a second example of endosymbiosis in eukaryotes. Today, chloroplasts are the organelles responsible for photosynthesis in plants and different types of eukaryotic algae. Algae is a general term for different groups of eukaryotes from single-celled forms to macroalgae, such as kelp, that blur the lines between colonies of individuals and true multicellular organisms.

About a billion years ago, algae had become vital components of early marine ecosystems, eventually providing additional resources for the first animals over 542 million years ago. However, it took more than 500 million years from the time a eukaryote engulfed a cyanobacteria to the evolution of the first plants living on land. The transition to living on land presented several major hurtles. First, you had to worry about drying out, so ways of preventing water loss had to evolve. If you live in water, then gravity is not much of a problem, but if you're on land, then you have to reinforce your cells walls to support yourself. If you're in water, it's easy to absorb all the nutrients you need directly into your cells from the water. But if you live on land, you have to absorb nutrients and water from the soil. And lastly, you can only absorb nutrients from the ground, so you have to evolve mechanisms of moving material between the leaves and roots of plants.

After four billion years of the Earth's surface being completely devoid of vegetation, the first plants began to colonize the land perhaps some 470 million years ago in a time period known as the Ordovician. At first plants were small and could only live near water or moist environments along streams and lakes. Based on the fossil record, molecular data, and common features of all modern plants, they most likely evolved from a single algae ancestor that strongly resembles modern day *Chara*, a resident of freshwater lakes and streams.

What Makes a Plant Unique?

Before getting started, I want to point out that at the cellular level all eukaryotes, including animal, plants and algae are fairly similar. They possess mitochondria for aerobic respiration. Meaning they use oxygen to efficiently produce ATP just like animals. They also have a cell nucleus and tiny membrane-bound organelles that compartmentalize the functions of the cell. Even though plants have mitochondria and use oxygen, their rate of photosynthesis, during the day at least, is higher than their rate of oxygen consumption from respiration. At night, plants plants continue to use oxygen and give off carbon dioxide, just like an animal.

Plants are autotrophic organisms, they are able to use the energy in sunlight to fix carbon dioxide to make all the organic molecules they require. Although, there are a few plants that have actually lost their photosynthetic ability. As a result, plants form the base of almost all terrestrial ecosystems, meaning lots of animals eat plants. As we will learn later, they are not without their defenses! They have cell walls made of cellulose to maintain shape and rigidity. To reinforce their cell walls, plants evolved strong proteins called lignin, making the cell walls even more rigid, which has allowed plants to grow into large trees.

Plants also have unique tissues and organs, which separate them from other types of alge. The most obvious organs are the leaves, an adaptation to increase surface area for photosynthesis. Their upper surface is typically waxy to prevent water loss. The underside of leaves have tiny openings called stomata that the plants can open or close to control water loss. Roots anchor plants to the ground and are used to absorb nutrients and water. A vascular system connects the roots to the leaves in plants.

Another defining feature of true plants is a unique life cycle where they alternate between multicellular haploid and diploid forms, in what is a called an alternation of generations. To understand alternation of generation in plants, it can be helpful to review our life cycle. Recall that humans produce single-celled haploid gametes (sperm and egg) each carrying a single copy of a genome on 23 chromosomes. The sperm and egg fuse to form a zygote, which will grow into a new diploid organism carrying two copies of the genome with 23 pairs of chromosomes, or 46 chromosomes total. Our adult form is diploid and multicellular; however, we produce single-celled gametes that do not divide or grow into a multicellular organism.

Plants are different, they have a multicellular haploid form called the gametophyte (phyte means plantlike), which produces single-celled gametes that fuse to form the diploid sporophyte. When you look at a tree, or the grass in your lawn, those plants are the sporophytes. In most modern plants, the gametophyte has become almost microscopic. It is this alternation of generations with plants having two multicellular life-stages, the sporophyte and the gametophyte that makes plants a unique lineage of organisms, separating them from algae. Although in flowering plants, the gametophyte is very small, consisting of just a few cells.

The Evolution of Plants

It's not quite clear when the first plants colonized the land. Fossil records support a date of approximately 450 million years ago, with some evidence pushing that date back to 470 million years ago. These first plants strongly resemble modern day mosses known as bryophytes, small plants found living in moist environments. Similar to modern mosses, they lacked true leaves, roots, and a vascular system to move water and nutrients from one end of the plant to the other. They also lacked ways to prevent drying out in the atmosphere, restricting them to moist areas. Although these early mosses had cell walls made of cellulose, they lacked proteins called lignin so they were unable to grow very tall.

For nearly 30 to 50 million years, almost the same amount of time between the extinction of the dinosaurs and the evolution of humans, plants remained relatively small and limited to growing in wet or moist areas. By 420 million years ago, the first vascular plants had evolved that resemble modern-day ferns. Evolving a vascular system capable of transporting water and nutrients from the roots to distant parts of the plant was a major evolutionary break through that when combined with lignified cell walls and true roots, allowed plants to grow tall. With these new features, the first forests began to emerge and were well underway by 400 million years ago. For the first time in 4.2 billion years of Earth's history, the surface was becoming green with life. The spreading of plants on the surface also greatly reduced erosion, increasing the amount of soil, further helping the spread of plants and expanding terrestrial ecosystems in a positive feedback loop.

From 360 million years ago to about 300 million years ago, the climate was wet and warm. Carbon dioxide levels were approximately 2,000 ppm, more than 5 times higher than today, keeping the Earth rather warm. The continents had drifted near the equator, and large tropical swampy forests dominated by giant fern trees covered much of the Earth's surface. Paleontologists actually refer to this time period between 359 – 299 million years ago as the Carboniferous Period because much of our coal deposits were formed during this period of Earth's history. The wet, swampy conditions were conducive to coal formation by preventing oxygen from breaking down the organic matter and releasing it back to the atmosphere as carbon dioxide.

Over millions of years, the accumulation of coal deposits reduced carbon dioxide levels in the atmosphere causing the climate to become cooler and drier. By the end of the Carboniferous Period, the Earth entered an ice age as carbon dioxide levels dropped to about 250 ppm. In

Chapter 14, I will go into more details about how life has influenced the atmosphere and climate. As the Earth's climate cooled 300 million years ago, the first seed plants known as gymnosperms evolved. Gymnosperm means 'naked seed' because the seeds develop on the surface of the leaves, scales or cones. They were well adapted to the cooler drier environments because seeds and pollen have protective coverings to prevent desiccation. By 251 million years ago at the end of the Permian Period, gymnosperms dominated the Earth's surface. Even today, they form the bulk of our northern boreal forest, still dominating in cooler and drier environments.

During the late Mesozoic Era, a time when dinosaurs were in their heyday and the first birds were taking flight, plants evolved flowers. The arrival of flowering plants would have an enormous impact on the future of diversity. Flowers were a novel adaptation to attract pollinators to help spread their pollen. Conifers use the wind to spread their pollen; because the process is random, they must produce copious amounts of pollen that they release to the world around them. If you've ever lived near pine trees, you may be familiar with the yellow pollen that covers your car in the spring. Flowering plants produce less pollen, instead they invest more of their energy into flowers and nectar, a sugary snack to entice animals to visiting the plants. Many flowering plants also evolved elaborate fruits to aid in seed dispersal.

Modern Plants Are Very Diverse

There are about 300,000 species of plants in the world, including bryophytes, ferns and their allies, conifers, and flowering plants. Modern plants range in size from little more than a tiny leaf floating on water, to giant sequoia trees reaching over three hundred feet tall. They are found on every continent, including Antarctica, and have evolved to live in extreme environments from the very cold, to hot and dry deserts, some have even returned to water. A few species of plants have turned to carnivory, while others have become completely parasitic giving up on photosynthesis all together. I'll start with the primitive bryophytes.

Even today, small bryophytes, relics of the first plants from a time long gone, are still present and found on every continent, including Antarctica. They are called non-vascular plants because they lack a vascular system and they also lack leaves and roots. Surprisingly, there are more than 16,000 species of bryophytes living today. Bryophytes are actually a generic term for three distinct groups of non-vascular plants called hornworts, liverworts, and even a lineage called the bryophytes.

Most of the seedless vascular plants have gone extinct, but one group called the horsetails or *Equisetum*, can still be found growing along stream banks and other wet areas. They date back to the Devonian some 400 million years ago they were once the dominant understory vegetation. Ferns are also an ancient lineage of seedless vascular plants dating back to about 360 million years ago. Today, there about 10,000 species of ferns found on most continents, except Antarctica. Most are understory plants, but a few species are trees. They do not produce, flowers, seeds or pollen, instead they produce very small spores which can be seen developing in specialized structures on the underside of their leaves.

Gymnosperms were the first seed plants and producers of pollen, evolutionary adaptations to cooler and drier environments. In seed plants, the male gamete is protected by a tough coat forming the pollen. Gymnosperms lack flowers and rely on the wind for pollination, so they produce lots of pollen. Today, there about 1080 known species of gymnosperms in the world. The most common gymnosperm are the conifers, which produce seeds in their cones. Conifers include some very common plants including, pines, junipers, redwoods, spruces, firs, and cypress trees. The largest forests in the world are the northern boreal forests or the Taiga stretching across Europe, Asia, and North America. It is dominated by the cold hardy conifers, capable of living in the harsh environments.

Modern conifers may potentially hold records for the oldest and largest organisms on the planet. A candidate for the oldest living organism on the planet is a bristlecone pine that is estimated to be 4800 years old. It's difficult to determine the largest single living organism on the planet, but it may be a Sequoia in California nicknamed General Sherman. It is larger than 24 blue whales, or 40,000 people. That's more people than are found in most universities in the U.S. One of the most interesting Gymnosperms is a plant called *Welwitschia* found only in the Namibian desert in Africa. It has only two leaves that grow continuously throughout its life, which some estimates place over 2,000 years!

The evolution of flowering plants, or Angiosperms, is one of the greatest success stories in life, illustrating how one evolutionary adaptation creates new opportunities for additional adaptations. The arrival of flowering plants was a driving force in the further diversification of animals over the last 100 million years, as diversity gives rise to diversity. Today, plant diversity is overwhelmingly dominated by flowering plants. There are approximately 300,000 flowering plants, with the greatest diversity in the tropics, compared to just over 1080 species of Gymnosperms, a group of plants that have been around twice as long as flowering plants.

Most flowering plants are placed into one of two large groups based on the number of leaves an embryonic plant produces. The monocots produce one embryonic leaf and eudicots produce two embryonic leaves. Another way to easily distinguish the two groups is by observing the patterns of veins in the leaves. Monocots include plants that often have parallel veins in their leaves. Examples are grasses, including bamboo, corn, wheat, and rye. Some monocots have beautiful flowers such as orchids and lilies. Monocots are typically not woody, lacking the ability to grow into large trees. The eudicots are easily recognized by the branching pattern of veins in their leaves. Eudicots include broad-leaf trees, bushes, and many of the wildflowers we are familiar with including, roses, peas, and sunflowers.

Today, flowering plants are found on every continent, including small grasses found in Antarctica. In the tropics and throughout much of the middle latitudes, broad-leaf trees dominate the forests. Broad-leaf trees in the tropics keep their leaves year-round to take advantage of the abundant sunshine. Although, they will replace them throughout the year. In temperate regions, where winter temperatures dip below freezing, many broad-leaf trees are deciduous, dropping their leaves in the fall, only to regrow them in the spring once temperatures warm.

In the American deserts, cactus plants have evolved several features to survive the arid conditions. Their leaves have been reduced to sharp spines providing protection to the plant while at the same time preventing water loss. As a result, cactus leaves are not photosynthetic, but the bark is, giving cactus their overall green appearance. Cactus are only found in the New World; although, there is one exception to this, a single species is found in Africa. However, cactus are not the only drought tolerant species. In Africa, a genus of plants called *Euphorbia* has also evolved reduced leaves to prevent water loss and larger stems capable of photosynthesis and holding water.

In contrast to cacti, some flowering plants have actually returned to aquatic and marine environments. Two examples are turtle grass (*Thalassia testudinum*) , a common plant in the Gulf of Mexico and eel grass *(Vallisneria americana)* in eastern streams, both resemble terrestrial grasses but are unrelated. Their similar form is an example of convergent evolution when two species evolve to look very similar because natural selection was operating in similar environmental conditions. In this case, a flat grass-like leaf is beneficial for maximizing surface area while reducing drag in the water.

Within the Cenozoic Era of the last 65.5 million years, the grasses have evolved to become one of the dominant terrestrial plants. Today, there are about 12,000 species of grasses covering nearly 20% of the terrestrial surface. You can find grasses living in parts of Antarctica, wetlands, forests, and tundra, not to mention that grasses are the dominant plants in tropical savannas and temperate grasslands. Grasses are also one of the most important plants for humans and account for much of our food production. Some of the most important grasses for agriculture include sorghum, rice, wheat, and corn. Rice alone may account for 1/5 of all the calories consumed by humans.

Modern flowering plants have many diverse lifestyles; the vast majority are only autotrophic. However, some plants have evolved "alternative" lifestyles. Mistletoe, the green plant you may have seen growing on the limbs deciduous trees in the winter, is actually a parasitic plant that sends its roots into the branches of other trees robbing them of nutrients. However, many species of mistletoe are still capable of photosynthesis, any plant that is green is photosynthetic. Not only is mistletoe a parasite, it's quite poisonous producing a chemical called phoratoxin, and eating it could cause nausea, diarrhea, and blurred vision. But, let's not forget that it may have medicinal properties as well, including potential cures for cancer

Growing on the forest floor in deciduous forest of the eastern US are small white plants called Indian pipes (*Monotropa uniflora*). They are a parasitic plant tapping into the roots of beech trees (*Fagus grandiflora*) in the eastern US, allowing them to grow in the very dark understory of forests where there isn't enough light for photosynthesis. It is white because it lacks the green pigment chlorophyll and therefore has totally lost the ability to conduct photosynthesis.

Some flowering plant species that live in very nutrient poor soils are carnivorous. They have evolved modified leaves to attract and catch unsuspecting animals to make a meal out of them. Perhaps the best known example is the Venus flytrap (*Dionaea muscipula*), a resident of swamps in North and South Carolina. Insects are attracted to the bright red inside of the highly modified leaves that are capable of rapidly closing. If an insect lands on the leaf and bends two or more of the "trigger hairs" the two leaves rapidly close catching the insect. Additional movement by the prey causes the leaf to seal off the insect and release digestive enzymes.

Pitcher plants also use modified leaves to catch insects. In this case, their leaves are shaped into a pitcher capable of holding water. To help trap the insects, the inside of the leaves forming the pitcher has small hairs pointing downward making it difficult for insects to land on the sides. Eventually, the exhausted animal falls into the water where it is slowly digested and its nutrients are absorbed by the plant. Also found alongside pitcher plants are Sundew (*Drosera* sp.) plants. Their modified leaves exude a sticky sap that catches small flying insects. Once an insect is caught, the leaf rolls up to digest the animal. Often seen floating in ponds are bladderworts (*Utricularia* sp.), a plant with one of the most complex traps among plants. They have modified leaves that form a bladder underwater and then actively pump the water out forming a vacuum. When a small animal touches the "trigger hairs", the trap quickly opens and sucks in the prey. Once inside the bladder, the plant releases enzymes to digest the prey.

In addition to flowers for attracting pollinators, flowering plants evolved fruits, a seed-bearing structure formed from parts of the flower. Fruits aid in seed dispersal, and like flowers, they too have become specialized in their modes of dispersal. Fruits come in a very wide variety, ranging from raspberries, to apples, oranges to very large coconuts. Technically, many items we eat including nuts, corn, wheat grains, tomatoes, chili peppers, rice, and beans are actually fruits.

Some plant species have coevolved to form symbiotic relationships with animals where the fruit provides nutrition to the animal and, in turn, the seeds are dispersed from the parent. If you've ever noticed more plants growing along a fence-row, it is from birds eating fruits and excreting seeds when perched on the fence. In fact, the seeds of raspberries and blackberries won't germinate unless they pass through the digestive tract of a bird. If you've ever picked up a sandspur or cocklebur, those are seeds that have evolved to get caught on mammal fur for dispersal. Some seeds such as dandelions and maple seeds are small and dispersed by the wind.

Plant and Animal Diversity Are Connected

Plant and animal diversity are closely tied to each other. There are an estimated 300,000 species of flowering plants, the end result of 120 million years of evolution with animals. The rate of diversification for the past 542 million years has not been constant, but has undergone several spurts in of rapid diversification, including the current period of high diversity because of flowring plants. Plants not only provide nutrients and energy to animals, but plants also provide structure to ecosystems, creating additional habitats for animals. In addition to creating structure, animal diversity has paralleled the diversity of flowering plants as both flowers along with their fruits and seeds have provided additional opportunities for diversification.

Every flower has an evolutionary history; it is the end result of millions of years of natural selection that has honed its ability to efficiently spread its pollen, often by attracting a pollinator. Sometimes, flowers have become incredibly specialized as they evolved to attract only one particular species of pollinator, while other flowers have evolved to attract as many species as possible. The end result is that flowering plants have greatly diversified into a myriad of colors, shapes, and sizes to attract pollinators. For example, large white flowers opening at night attract bats and nocturnal moths. Bright yellow sunflowers attract a wide variety of insects including bees, butterflies, and flies. Additionally, their seeds attract birds to aid in their dispersal away from the parent.

Many flowers are quite aromatic, producing pleasant smelling odors to attract many types of insects. Even humans enjoy the sweet smell of roses. However, not all flowers have a pleasant aroma like sweet roses. In southeastern Asia you may be able to find the world's largest flower, at least by weight. Except you wouldn't want to smell it because it has the aroma of rotting flesh. It is commonly called the "corpse flower" and its stinking flower attracts flies to serve as it pollinators. This large flower is in the genus *Rafflesia* and is a parasite of a particular vine, so there isn't much of a plant to look at when it is not in bloom. Ironically, the largest flower by size is the corpse lily *Amorphophallus titanum* of Sumatra. It is very rare, but its flower can grow up to 12 feet tall and weigh about 200 pounds.

In my opinion, the orchids produce some of the most beautiful flowers in the world. In many cases, a species of orchid has become so specialized, it can only attract be pollinated by one specific species of insect or bird.

This happens through coevolution when two or more species mutually interact over time driving the other's evolution. A classic example of coevolution is the hammer orchid (*Drakaea glyptodon*) that has evolved to be pollinated only by one species of wasp. To attract a male wasp, the flower resembles the female, but it also emits a pheromone that mimics the female. The male wasp will land on the flower and attempt a copulation, which will invariably fail. Fortunately for the plant, the wasp gets a full dose of pollen to spread to the next flower. There are about two thousand species of orchids that include some interesting and beautiful species including the Monkey Face Orchid, Moth Orchid, Bee Orchid, Flying Duck Orchid, Tiger Face Moon Orchid, Dove Orchid, and White Fringe Orchid.

Coevolution is not limited to flowers and their pollinators. Another example of coevolution between plants and animals are the acacia ants (*Pseudomerymex*) that live on bullhorn acacia trees (*Acacia cornigera*) throughout Central America. The tree has large hollow thorns where the ants live and the leaves produce a sugary drink supplied to the ants. In return, the ants fend off any herbivores that may try to eat the plant.

Tropical rainforests are the heart plant diversity due to rapid rates of evolution driven by competition, or symbiotic relationships, or for other reasons not yet known. In the Amazon Rain forest, there may be upwards of 300 species of trees per acre, and approximately 1000 flowering plant species. In addition to their diversity, trees and other plants also add structure to communities creating additional habitat that can be exploited by plants and animals. In fact, most of the diversity in rain forests are located in the canopy.

The forest canopy is comprised of the upper parts of all the trees including the trunk, stems, and leaves. The branches of large trees become covered themselves in plants, including bromeliads and orchids, all providing additional resources numerous animals. It has been estimated that an acre of lowland Amazonian rain forest may have 30 billion arthropods. Another study in Ecuador discovered over 100,000 species of plants and animals per acre in the canopy alone.

Plants are Chemical Factories

Plants are often passed over as boring compared to the more charismatic animals. But, we shouldn't underestimate plants, it turns out that plants are not as passive or defenseless as once thought. They make up for their sessile lifestyle by producing thousands, if not millions of different chemicals for various purposes including defense against herbivory, communication with other trees, and attracting pollinators to aid pollination. The sweet smell of roses, the savory taste of garlic, and the heat of chili peppers are all sensations caused by chemicals known as secondary metabolites created by plants. Recall that metabolism is the sum of all chemical reactions in an organism, so a primary metabolite in a plant would be sugars. Chemicals made after the formation of sugars would be made second, hence the name secondary metabolite. They are also called phytochemicals, because they are made by plants. Let's take a closer look at some of these chemicals that are both beneficial and harmful.

If you've ever improved the taste of your favorite foods by adding herbs and spices including, rosemary, thyme, garlic, cinnamon, oregano, pepper, *etc.*, then you know that plants produce many aromas and taste that we enjoy. Most of these compounds that we use to add flavor to our foods are actually made by plants to fend off insect herbivores. What is harmful or taste bad to an insect may actually be beneficial to humans. Indeed, scientific research actually backs up this claim. In one large study, scientists showed that people who cooked with a variety of herbs and spices actually had lower rates of disease and parasite infections.

Caffeine and nicotine are two examples of compounds produced by plants to prevent herbivory from insects. They are both stimulants and addictive, although caffeine not as much as nicotine. Scientific studies have shown that small amounts of caffeine from tea and coffee can improve cognitive function and reaction times. However, too much caffeine can cause health problems including rapid heart rate, loss of sleep, and anxiety in some individuals.

Living in New Mexico, most of us enjoy adding green chili peppers to just about everything from pizza to breakfast burritos to cheeseburgers. In fact, I would consider green chili to be the plant equivalent to bacon for its versatility in its ability to improve the flavor of so many foods. Some industrious cooks go so far as to roast chili peppers stuffed and wrapped in bacon! The active ingredient in chili peppers making them hot is called capsaicin, which is an irritant that we perceive as heat. Contrary to popular belief, the seeds of peppers do not have capsaicin, but the areas around the seeds do. Birds spread pepper seeds because the seeds pass through the digestive tract unharmed. Mammals, on the other hand, have large teeth that grind the peppers, destroying the seeds. Therefore, it's most likely that peppers evolved capsaicin to prevent herbivory from mammals because birds are unaffected by capsaicin.

Although, capsaicin is an irritant, it can be applied to the skin to relieve minor muscle aches and pains. Claims have been made that it may prevent cancer; however, there is not enough data to support this claim yet. A very large corollary study did show that mortality was inversely related to spicy food consumption, but the reasons for the lower death rate are not known and is an area of current research.

Over 50% of our medicines come from plants. For thousands of years, people have turned to plants for remedies to cure all kinds of ailments including nausea, toothaches, diarrhea, aches and pains. Perhaps one of the most widely used medicines from plants is called Acetylsalicylic acid, which is better known as the active ingredient in aspirin. This chemical was first isolated in the 1850s from the bark of the common willow tree (*Salix*), although the therapeutic properties of willow trees have been known for over 2400 years, as the Greek philosopher Hippocrates prescribed it for headaches. In California, grows the endangered Pacific yew tree (*Taxus brevifolia*) whose bark contains an anticancer chemical called paclitaxel. It works by preventing cells from dividing. Researchers at Florida State University were able to synthesize it to be used as an effective cancer treatment.

Primates, including humans, have lost the ability to produce vitamin C because we obtain it from the foods we eat. In recent years some plants have earned the label as a super food because they are packed with vitamins and minerals that are required to keep us healthy. There is no consensus on the exact ranking of foods in their nutrition, but there are some good guidelines. Green leafy vegetables including kale, spinach, and broccoli are considered to be very healthy and are recommended as part of a healthy diet. Other healthy foods include blueberries, garlic, and peppers. The idea is to eat a varied diet with multiple colors of fruits and vegetables to ensure good health. For example, red bell peppers actually contain more Vitamin C than oranges and other citrus fruits.

While many plants are beneficial to humans, some are very poisonous. A poison is any substance that can cause harm to an organism and a toxin is more narrowly defined as a poison produced by an organism. Poisonous plants and animals produce toxins, this means that harmful substances such as heavy metals, commercial insecticides, and plastics are poisons, but they are not toxins. Plants produce their toxins to prevent disease, fungal infections, herbivory from mammals and insects, and to prevent competition from other plants. One famous poisonous plant is Hemlock in the genus *Conium*, which is neurotoxic. In 399 BC, the Greek Philosopher Socrates was sentenced to death; to carry out the sentence, he drank a cup of hemlock tea and died within a half hour.

The deadliest plant in the world is named *Nicotiana rustica*, commonly known as tobacco. It produces the powerful stimulant nicotine, which is also incredibly addictive. Nicotine is an opiate related to morphine and caffeine, all made by plants to prevent herbivory from insects. Every year, 5-6 million people die of preventable deaths, the direct result of smoking tobacco. No other plant causes more death than tobacco.

It turns out that plants are far from passive organisms, they are capable of rapid responses to herbivory, quickly deploying their chemical defenses. In a very surprising discovery in the 1980s, it was found that plants also produce chemicals to communicate with each other, they can actually talk to each other. Rather than using sounds waves like animals, they send out chemical signals warning nearby trees of danger. All organisms have evolved to conserve their resources, and it takes a lot of energy and resources to produce chemicals to prevent herbivory. If you're not being eaten, why produce the chemicals to ward off a particular insect. It turns out that if a plant is getting grazed upon by an insect or mammal, it begins to produce compounds that will make it taste bitter, so the animal moves to the next plant. But, the plant has also sent a chemical message through the air to other plants. Upon receiving the signal, the nearby plants begin to produce chemicals making their leaves bitter to the grazing animals.

Part 2: Fungi
Origins of Fungi

The evolutionary origin of fungi has been difficult to determine because they don't fossilize well. To further complicate the story, the fossils we do have can be difficult to interpret, and DNA evidence has been varied. Piecing together all the available evidence, the origins of fungi began somewhere between 700 million and 1.5 billion years ago when their ancestors split from other eukaryotes. The last common ancestor of fungi and animals was probably unicellular, and multicellularity evolved independently in each group. In fact, some fungi, including baker's yeast have remained unicellular.

Similar to plants and animals, fungi originated in water and colonized the Earth's surface later. However, fungi may have been among the first organisms along with plants to colonize the land around 450 million years ago. Evidence from the fossil record indicates that fungi may have quickly formed symbiotic relationships with the earliest plants, helping them to survive on land by improving their ability to obtain nutrients from the primitive soils. By the Devonian 400 million years ago, fungi had become common in the fossil record and were an integral part of terrestrial ecosystems. Most of the modern groups of fungi were present by the end of the Carboniferous 299 million years ago. The first mushroom-forming fungi were present by the mid-Cretaceous, about 90 million years ago.

Fungi Exhibit Diverse Lifestyles

Fungi, like plants and animals are comprised of eukaryotic cells. Independently of plants, fungi evolved a cell wall made of chitin that is even more rigid than cellulose. Historically, it was generally presumed that fungi were more closely related to plants than to animals because of their superficially plant-like characteristics including a cell wall, sessile life style, root-like structures, and plant-like growth. However, the relationship to animals is supported by molecular data and the fact that they are both multicellular heterotrophs lacking chloroplasts. Heterotrophic organisms must acquire energy by breaking down organic molecules they obtain from their environment. The hyphae of fungi resemble plant roots in that they both absorb materials from their environment. Hyphae are very small and produce numerous enzymes to break down material so they can absorb it.

The total diversity of fungi is not well known, there are an estimated 100,000 species that have been described so far. Although, estimates of total diversity range as high as one million species. We can broadly divide fungi into two groups; macroscopic fungi that are easy to see, and microscopic fungi that require a microscopic to be observed. Many familiar groups are macroscopic, including the familiar mushroom-room forming fungi, most of which are decomposers. Lichens are an interesting group of fungi that forms a symbiotic relationship with algae, allowing them to survive where few other living organisms can. However, most fungi are inconspicuous due to their small size and cryptic nature. Microscopic fungi are often pathogenic, meaning they can cause diseases and discomfort, including common ailments like athlete's foot.

Decomposers: More than once I've cultivated some pesky mold that seemed to appear out of nowhere on my leftover lunch that stayed too long in our refrigerator. Although, I often become sad having lost my leftovers to some fuzzy looking mold, but I can't help wonder how it got there? What type of mold is it? Is it unique to bread or the cheese, or the fruits? And how is it making a living on my left-over lunch? Black bread mold is common on bread and is usually caused by the species

Rhizopus stolonifera. Because fungi are heterotrophic, the hyphae excrete chemicals into their environment to break down and absorb organic molecules required for their survival. Fungi also wage chemical warfare by producing chemicals to protect themselves from bacterial infections and to prevent competition from other molds growing near them. Unfortunately, if your bread becomes moldy, it's best to not eat it because many of these chemicals are toxic to us and the hyphae may be growing throughout your bread.

Before we all become mold-phobic, don't forget that many of our favorite cheeses are the result of certain types of mold that aren't bad for us. Blue cheese is made by adding a mold in the genus *Penicillium* to the final product providing the cheese with its distinctive smell and blueish color. If you recognize the name of that mold, it is because our first wide-spread antibiotic used to treat bacterial infections was isolated from a compound made by a similar species of mold in the same genus! Pathogenic bacteria can be harmful to plants, animals, and fungi. As a result, molds in the genus *Penicillium* protect themselves from bacterial infections by producing antibiotics that prevent bacteria from forming a cell wall, killing the bacteria.

When walking through a forest, on campus or, even in your backyard, you may have noticed mushrooms popping up in circles or in lines, especially after rain. The mushrooms that we see above the ground are actually a small part of the whole organism. It is only the fruiting body producing tens of thousands, and in some cases, billions of spores to be spread by the wind. Below ground are the hyphae, which secrete chemicals used to breaking down dead plant and animal material so that it can be easily absorbed. In some cases, the total below ground mass of some fungi are thought to be massive, and may actually be among the largest organisms on the planet. One example is located in the Blue Mountains of Oregon where a honey fungus (*Armillaria ostoyae*) is thought to be 2.4 miles across and may be between 1900 and 8650 years old. The evidence used to support that it is one organism lies in the observation that all the individuals are genetically identical and recognize each other just like the cells in our own body recognize each other.

As a decomposer, many of our larger fungi are important for breaking down organic matter so that the nutrients and minerals can be recycled and used again. Without the action of decomposers, much of the world's nutrients would soon be locked up in dead plants and animals. Part of the reason why fungi are so diverse is easily described with the simple statement: Diversity gives rise to diversity. As we saw with the evolution and subsequent diversification of flowering plants, insects also quickly diversified. Fungi followed the same pattern as plants, they have greatly diversified mirroring plant and animal diversity. New species of fungi have emerged as they have coevolved with plants and animals. In fact, whenever you observe a unique looking fungus, it may only be found on a certain type of tree, tree root, cheese, or animal.

Mutualists: Fungi maybe well known as the decomposers of ecosystems, but many types also form mutualistic relationships, mostly with plants and algae, and in some cases they are even farmed by ants! *Mycorrhizal* fungi form close relationships with roots of many plants species improving the growth of the plant. This makes sense because fungi produce many types of powerful enzymes capable of break down organic material, or even rocks, liberating essential nutrients and minerals making them easier for plants to absorb through their roots. Some fungi form such a close relationship with plants that their hyphae even enter and live within the root's cells.

If you've ever been hiking in the woods, you've probably noticed lichens growing on the trees, on the rocks, or even on the ground. Lichens are incredibly hardy due in part to their symbiotic relationships with algae. The relationship is mutually beneficial. To benefit the algae, the lichens secrete enzymes capable of slowly breaking down rocks liberating minerals required for the algae to grow grow. The tough cell wall of the lichens can prevent desiccation while other chemicals can make them toxic preventing herbivory. The algae benefit the lichens, by providing them with a source of organic molecules and energy.

Humans are not the first farmers. That distinction goes to ants, those venerable insects that appear in just about every terrestrial ecosystem.

Although, not every species of ants are farmers, this title goes to the leafcutter ants of Central and South America. In all, there are 47 species of these ants that harvest plant material to grow type of fungus called *Mycelium*. Next to humans, leaf cutter ants exhibit the most complex societies in the animal kingdom. Colonies can reach upwards of 8 million individuals and cover several thousand square feet. The relationship between the ants and their fungi is incredibly well developed and actually involves a third player. Growing on the ants is a species of bacteria that secrete antimicrobial compounds to prevent undesirable bacteria from causing problems for the fungi. Recall that many types of plants produce toxic chemicals to prevent herbivory. If a leaf turns out to be toxic to the fungus, it will send out chemical signals to the ants communicating to them not to collect that particular type of leaf.

Parasites: Can a fungus really be a parasite? The answer to this is: yes, but it's complicated. Parasites have a very close symbiotic relationship with their host. Rather than being mutualistic where both partners benefit from the arrangement, parasites benefit at the expense of their host. As we will learn in the next chapter, many parasites have complex life cycles and are capable of altering the behavior of their animal host to improve their chances of transmission. The same is true of the *Cordyceps* fungus, which turns many insects into a living zombie. When the *Cordyceps* fungus infects an insect, it will eventually kill the insect. Once an insect becomes infected with the fungus, it will alter its behavior by causing it to move to an open area and latch on to a branch or leaf only to die in place. Once the unlucky insect dies, the fungus begins to act more fungus-like, slowly decomposing the animal to acquire nutrients. After a short time, it will begin to grow out of the insect and release its spores to infect other insects. There are about 400 known species of *Cordyceps* fungi, each one infecting a different species of insect. As you can see, the *Cordyceps* fungus acts like a parasite because it infects a live organism and alters its behavior to improve its chances of infecting another insect. Some observations have shown that ants infected with the fungus will actually climb a branch above their own colony. If an infected ant is spotted by another individual ant from that colony, the infected ant will immediately be carried away from the colony.

Disease Causing Fungi

Along with prokaryotes, fungi are also part of our microbiome, although they may not be as beneficial as many of the prokaryotes inhabiting us. Most microscopic fungi living on our skin largely go unnoticed. In fact, it wasn't until improvements in our ability to sequence tiny fragments of DNA combined with the curiosity of a few scientists who went around taking swabs from peoples' skin that it became apparent there were numerous types of skin fungi. In fact, the average person may harbor between 30 and 60 different types of fungi at any one time. One of the most common types of fungal infections is athlete's foot, which causes dry flaky skin and irritation on feet. Other types of fungal infections cause dry and brittle toenails, which can be difficult to treat.

The rapid decline of amphibians in the last few decades has been one of the most tragic losses of an entire lineage of vertebrates in modern times. As we will learn in chapter 14, there are many reasons driving amphibian declines worldwide. But, one confirmed cause has been the spread the Chytrid fungus throughout the Central and South America. The Chytrid fungus has devastated entire amphibian communities throughout the tropics over the last several decades. Entire populations of frogs and toads have been wiped out within a few months after the arrival of the fungus.

In 2008, I travelled to a remote region of Panama with herpetologists from the University of New Mexico to find rare reptiles and amphibians. After my first couple of nights, I was happy to have found 10 species of frogs along a one-mile trail that meandered through the tropical rainforest. Over the course of a week of intense herping (that's looking for reptiles and amphibians) we managed to find 20 species total. Near the end of the trip, I learned from another colleague who had been there 1.5 years earlier that there were once almost 90 species of frogs and toads along the same trail. In a very short time, the Chytrid fungus spread to this area wiping out nearly 80% of the amphibian diversity in less than a year. Surveys in subsequent years have not found most of those missing amphibians.

Amphibians are highly susceptible to fungal infections on their skin because they use it to breathe and regulate electrolytes when they are in the water. The Chytrid fungus infects their skin rendering them unable to regulate their electrolytes, resulting in a quick death for the animal. Fortunately, the Chytrid fungus is not fatal to every species of amphibian.

Our food crops can be infected with a black colored fungus called ergot (*Claviceps*), which produces several types of chemicals that are toxic to humans. The typical symptoms include nausea and vomiting and in some cases hallucination. Ergot has been documented infecting an entire town's wheat supply. In fact, the Salem Witch Trials, when the town "went mad", may have been caused by a large outbreak of ergot causing numerous people to become sick and begin hallucinating.

Yeast and Civilization

Civilization has its roots in the agricultural revolution that allowed people to transition from a wandering hunter-gatherer life-style to taking up more permanent residence. Once people were free of having to find food everyday, they were able to pursuit other interest. Within a short time, there were farmers, builders, poets, philosophers, scientists, teachers, politicians, and lawyers (that list is not meant to be in any particular order). Why did people abandon the hunter-gatherer lifestyle, what brought about the agricultural revolution?

Numerous hypotheses have been put forth to explain why our ancestors settled down and began farming. The fact that the agricultural revolution started about the time of the last ice age is not lost on anthropologists. Ever since the last ice age, humanity has greatly benefited from some of the most stable climates in the history of our species. As you might guess, the stable climate may have been a major factor in starting agriculture because you could predict when to plant and harvest your crops.

Another hypothesis to explain the start of agriculture centers on the evidence that our ancestors began planting and harvesting crops to brew alcoholic beverages including beer, wine, or mead. Ancient hieroglyphics from Egypt and other early civilizations clearly depict people drinking fermented beverages. Beer was used by the Egyptians to pay workers building the pyramids a wage of one gallon of beer per day. Other ancient civilizations had it written in their laws that beer could serve as payment.

There may have been several benefits of drinking beer in ancient times. Diarrhea caused by various waterborne pathogens including *Giardia* and *Cholera* has been and continues to be a major cause of death in humans. Brewing beer or other alcoholic beverages would have reduced the risk of getting sick from drinking contaminated water.

Alcoholic beverages are typically made fermenting some type of plant material by yeasts, a single-celled fungi capable of breaking-down the sugars in many types of plants. In the absence of oxygen, they produce ethanol as a by-product of the fermentation process. Ethanol found in the beer would have killed many of the pathogens making it safer to drink than normal water. Additionally, beer would have been a source of vitamins and carbohydrates for energy. Before you decide to give up on water and turn to a diet of beer, you should know that ancient beer was about 2-3% ethanol, compared to the much higher alcohol content upwards of 8% found in modern beers.

The other reason why beer may have been important for the rise of civilization lies in its ability to reduce social anxiety among people allowing them to more freely mingle and speak their thoughts. Basically, beer served as social lubrication and was an important part of feasts and celebrations that promoted social networking required in a civilization. But the main question that we may never know the answer to is, what prompted the first person to drink water with rotting wheat or fruit in it? I suppose it was some guys daring another guy to drink something gross.

Chapter 9

Most Animals Are Invertebrates

To a first approximation,
Most animals are arthropods.

Introduction

If an alien race were to visit our planet and randomly sample the diversity of animals, they would quickly conclude that most animals are arthropods, an invertebrate. Traditionally animals have been divided into two broad groups; the vertebrates, which are animals with backbones, and the invertebrates, which are animals without backbones. From an evolutionary, point of view, this distinction is some somewhat arbitrary as starfish and sand dollars are more closely related to vertebrates than they are to other invertebrates.

Many of the animal phyla are wormlike in appearance. Some, such as the peanut worms are rarely encountered, while nematodes (roundworms) are nearly ubiquitous and found in large numbers, but most are small and go relatively unnoticed. Some worms are parasites, and superficially the appear quite simple, yet have evolved extraordinary life histories as they go between different host organisms completing complex life cycles. Sometimes these parasites even manipulate the behavior of their host, in some cases dooming them to die.

As you learn about animal diversity, keep in mind that all animals share a single common ancestor that lived about 600-700 million years ago. As a result, all animals have traits in common, inherited from that last common ancestor. Yet, there is incredible diversity of animals based on modification of these common characteristics.

Invertebrates are found practically everywhere. Small crustaceans and snails are found at the bottom of the ocean at depths exceeding 35,000 feet, while small spiders float miles high in the atmosphere using their webs to catch the wind. I start this chapter with a brief discussion of the evolution and diversification of animals. But, animals are so diverse, it is impossible to cover the entirety of animal diversity in a single chapter. Therefore, I'm going to focus on just a few of the major groups of invertebrates to provide a glimpse of the stunning diversity of animals found in this world.

The Rise and Diversification of Animals

Over time, life evolves, adapts and another theme in evolution is that diversity gives rise to diversity. If the environment changes or new opportunities become available, life finds a way to exploit the new opportunities. But, why did life only exist as simple single-celled organisms for over 3 billion years? What caused the relatively rapid evolution of multicellular animals from single-celled eukaryotic cells? The answers to these questions may be difficult to find, but hard questions don't stop scientists from trying to answer to them.

Part of the problem lies in the fact that it is difficult to find the microscopic fossils of single-celled organisms in rocks hundreds or even billions of years old. Another difficulty arises from the geological activity of the Earth itself. Continents move around, mountains are uplifted and weathered away, rocks containing ancient fossils are slowly eroded, removing any indication they ever existed. The further we look back in time, the less evidence we have of early life. Despite all these difficulties, we have made some great fossil discoveries and developed several theories to explain the relatively sudden appearance of animals about 542 million years ago in an event affectionately called the Cambrian Explosion.

Darwin, along with many other naturalists of the 1800s, noticed that fossils of animals were abundant from Cambrian Period rock layers and above, but there were not any visible fossils of animals below these rocks. It's as if animals suddenly appeared in the fossil record fully formed. In Darwin's time, scientists lacked the tools or know-how to find microscopic fossils or to accurately age rocks. It took the discovery of radioactivity in the early 1900s and another 100 years of hard work to accurately date rock layers around the world, placing a timeline to the fossil record.

The technique used to date rocks is based on the steady and predictable rate of radioactive decay of certain isotopes. Recall that all elments have isotopes, which are formed by varying numbers of neutrons in the nucleus. For example, carbon always has 6 protons in the nucleus, but it can have 6, 7, or 8 neutrons making carbon-12, carbon-13, and carbon-14. Some ratios of protons to neutrons are stable while others are not and decay into daughter elements with more stable combinations of protons and neutrons. For example, carbon-14 (6 protons 8 neutrons) is unstable and decays into nitrogen-14, a very stable element. Luckily for scientists, these radioactive isotopes decay into daughter elements at a very predictable rate called the half-life. By using the ratio of radioisotopes to the daughter elements, an accurate age can be determined for when a rock was formed. Most people are aware of carbon-14, but every element contains radioactive isotopes. For ancient rocks, scientists may use multiple elements including uranium and potassium to determine their age.

Once scientists began to determine the age of the earliest animal fossils, the next question focused on the causes driving the rapid rise of animals in the fossil record. Below are some of the leading hypothesis used to explain the appearance of animals at the start of the Cambrian time period. Keep in mind, that the actual answer may be a combination of these factors.

Hypothesis 1: The Rise in Atmospheric Free Oxygen

Every rock holds clues to how and when it was formed. As we analyze the mineral composition and ratio of isotopes in rocks going further back in time, a picture emerges of the ancient Earth that lacked free oxygen (O_2) in the atmosphere for its first 2.5 billion years. Over time, through the constant action of photosynthetic organisms, oxygen levels slowly increased. By 2 billion years ago, oxygen levels were about 2-3% of the atmosphere, enough for bacteria to evolve the ability to use oxygen for aerobic respiration. Then about 600-550 million years ago, oxygen levels rose to about 12-15% of the Earth's atmosphere. In a very short time, geologically speaking, oxygen was six times more abundant than it had ever been.

Free oxygen in the atmosphere energizes life, it allows cells to extract 15 times more energy out of the same amount of food, thus enabling an active life style, a hallmark of animals. For this reason, the rise in atmospheric oxygen levels has been proposed as a major reason why animals were finally able to evolve. Without the oxygen in the atmosphere, the active life style of animals would not be possible. People who hike to the top of tall mountains have experienced sluggishness resulting from less oxygen at 12,000 feet or more in elevation.

Hypothesis 2: The Evolution of *Hox* Genes

Animals are complicated organisms, as an animal, we are a colony of approximately 10 trillion genetically identical cells working together for the common good of a single multicellular animal. Our cells may be genetically identical but, we have over two hundred different types of specialized tissues, including muscle, nervous, skin, *etc.*, to carry out the multiple functions of our bodies. Animals also have symmetry, which can be thought of as an organized body plan. The simplest body plant is radial symmetry making the animal appear circular. More complex animals, like ourselves, have bilateral symmetry where the animal has a front and back, top and bottom, and left and right sides.

In order for an animal to develop from a single-celled zygote into a multicellular animal with a body plan, requires the coordination of numerous genes. Genes have to be highly regulated, turned on at precisely the right time and then turned off again. To achieve this level of coordination in growth and development requires an entirely new type of master genes capable of coordinating growth and development at the genetic level.

The answer shared by all animals are the *Hox* genes. These are master regulatory genes used to coordinate the growth and development of animals from a single-celled zygote into an adult. They must have evolved early in the history of animals and have remained relatively unchanged, although some animals have more *Hox* genes than others. Several interesting experiments have shown how these genes can still function among different taxa that have been separated by hundreds of millions of years of evolution. For example, if you insert human *Hox* genes into mice, or even the more distantly related fruit flies, these genes will still produce fully functioning fruit flies and mice. It appears that development from a single celled zygote is a highly conserved process across all species of animals.

Mapping the position of *Hox* genes on chromosomes has revealed that they are organized in our genome based on their functions. Similar to fruit flies, we are segmented, a quick examination of our spine and ribs reveals a segmented body concealed by our skin. Each one of our vertebrae is numbered, indicating its position in our body. *Hox* genes line up with the segmentation in all animals, coordinating growth and development by turning genes on and off at the correct time. Understanding the evolution of *Hox* genes and how they work is a major area of current research because it could provide breakthroughs in growing lost limbs, or damaged organs.

Hypothesis 3: The Rise of the Predators

One of the favorite hypothesis among paleontologists to explain the sudden appearance of animals is the rise of the predators setting off an evolutionary arms race. The closest living relative to animals is a single-celled eukaryote called a *Choanoflagellate*. These tiny organisms may have faced the dim prospect of being eaten by other, larger predaceous eukaryotes. To avoid predation, several *Choanoflagellates* could join together forming a small colony that was too big to be eaten. Once a small colony was formed, some individual cells could specialize in certain tasks allowing the colony to pursue new resources, including eating other eukaryotes or other colonies of eukaryotes. And so an evolutionary arms race began, quickly driving the evolution of animals and their rapid diversification.

You could imagine an evolutionary arms race going something like this: prey evolve to be larger to avoid being eaten, predators evolve to be larger to go after the larger prey. The prey evolves appendages allowing them to move away from the predators, the predators evolve better fins and stronger muscles allowing them to swim faster, or become more agile. The prey evolves a hard outer shell to protect themselves from being eaten, the predator evolves teeth, jaws, or claws to break the shell. The predator evolves eyes to detect the prey, the prey evolves eyes to detect and avoid predators, and so it goes. In fact, eyes may have evolved independently in animals about 20 times.

My guess is that the rise of the animals was triggered by a combination of all three of these events occurring at about the same time, a perfect storm for evolution. With more free oxygen available, animals could become larger and more mobile fueling the evolutionary arms race. The presence of *Hox* genes allowed for multicellular life to quickly diversify. It requires only small changes in the timing of *Hox* genes to produce very large changes to the form of an animal. As a result, animal life quickly diversified into many different forms. By the end of the Cambrian Period 499 million years ago, all the modern groups of animals we are most familiar with, including arthropods, echinoderms, mollusks, cnidarian, and chordates, had all arrived and were diversifying.

What Makes an Animal Unique?

All modern species descended from ancestral species by accumulating modifications over time as they adapted to their environment. By studying different types of animals and comparing their similarities, we can piece together the characters shared by the last common ancestor of all animals. Taken together the similar features of animals can be used to define the kingdom Animalia as a unique group, different from all other organisms. Below is a description of the common features of animals, but it is not necessarily in any particular order.

In many introductory and high school biology courses, you may have suffered through the rite of passage of dissecting some fetal pig or feral cat. But, getting past some squeamishness of cutting open an animal to reveal its insides, you realize that the organs and anatomy is highly organized. It turns out that our organs and tissues arise from three different germ layers during development. In fact, all animals with bilateral symmetry have three germ layers, the endoderm, mesoderm, and ectoderm. The endoderm forms our digestive tract including our esophagus, lungs, stomach, liver, and the colon. Our mesoderm forms muscles, bone, connective tissue, fat tissue, and circulatory system including the heart. The ectoderm forms our skin, teeth, hair, and parts of our nervous system.

Animals also have symmetry that comes in two different types: radial symmetry and bilateral symmetry. Cnidarians are a primitive group of animals with radial symmetry and two germ layers (endoderm and ectoderm). Despite their simple body plan, cnidarians are a diverse and important group of animals including sea anemones, jellyfish, and corals. All other animals have bilateral symmetry; front (dorsal), back (ventral), left and right, top (anterior) and bottom (posterior) ends. Bilateral animals move through environment with their head first. As a result, many of their sense organs including, eyes, nose, ears, and antennae are located on the head.

Animals are able to actively move through their environment, whether using legs to walk, wings to fly, or fins to swim. Therefore, all animals have muscles allowing them to move. Some animals such as the barnacle give up their ability to move through the environment as adults, but they still have muscles allowing them to extend their legs to collect plankton passing by. To coordinate muscle movement, animals also possess a nervous system, another unique feature of animals that allows for rapid long-distant communication between cells.

Lastly, animals are heterotrophs that feed by ingestion. Animals are among the most active organisms on the planet. To power this active life style, animals require lots of energy and feeding by ingestion is a great way to obtain lots of energy. This one requirement is part of the reason behind the diversification of animals into so many species from herbivores to predators. The high energy demands of animals are one reason why they can't be totally photosynthetic similar to plants. However, if there is one rule in biology, it's that some animal, somewhere, has evolved some way of defying the general trends. As we will learn, some animals have lost their mouths while other animals are photosynthetic!

Practically every classification system includes sponges as animals. They are multicellular organisms with specialized cells and they feed by ingestion. It is quite clear, based on molecular data and the presence of certain cell types that sponges shared a common ancestor with all animals and are therefore closely related to them. In some ways, they represent the transition from colonies of eukaryotes with specialized cells to animals with specialized tissues and symmetry. However, because sponges lack true tissue layers, symmetry, muscle and nervous tissue, I consider them to be "para-animals" or sister to animals. But, not everyone agrees with me.

Cnidarians
Stinging Tentacles and Reef Builders

Cnidarians are often considered to be "primitive animals" due to their simplistic body plans and lack of a brain. They have radial symmetry, which is much like a circle, if you were to draw a line through the middle, the two halves would always be the same mirror images of each other. They only have two germ layers, an ectoderm and an endoderm and lack a head with advanced sense organs, although they can detect light. Their name comes from the term cnidocytes for their stinging tentacles, a character shared by all cnidarians.

I'm always a little hesitant to call animals primitive, Cnidarians have just as long as an evolutionary history as us humans. Despite remaining relatively simple, their body plan has worked well for them for 600 million years, so it's changed very little in all that time. They are a great example of natural selection acting to maintain a successful body plan. Today, Cnidarians remain a diverse group of animals with about 10,000 species in four major classes: the jellyfish (Scyphozoans), corals and anemones

(Anthozoans), sea wasp and box jellies (Cubozoans), and the Hydrozoans including the tiny hydra and the much larger Portuguese Man-O-War.

Most of us are familiar with the jellyfish in the class Scyphozoa, which means 'cup-like animal'. They are free-floating animals capable of swimming by simply contracting their bell. Many species possess light sensitive pigments, which are used to guide them through daily vertical migrations in the water, ascending from the depths towards the light of day. So far, about 200 species of jellyfish have been discovered throughout the world's oceans from the tropics to the poles.

The largest jellyfish is the Lion's mane Jellyfish (*Cyanea capillata*) found in cold arctic waters, it can reach 7.5 feet in diameter with its tentacles trailing nearly 40 feet behind it. Adult jellyfish of smaller species are a favorite of the leatherback sea turtle who sometimes eat plastic bags, mistaking them for a jellyfish. In recent years, overfishing has drastically reduced fish populations, which are natural predators of jellyfish. As a result, some populations of jellyfish are exploding, especially in the Sea of Japan. To make matters worse, the large numbers of jellyfish are eating the babies of the same fish being harvested, further exacerbating the spiraling decline in fish populations.

While many jellyfish can inflect annoying stings, most aren't deadly to humans unless you're allergic to them. This is not the case for another lineage of cnidarians called the sea wasp or box jellies (class: Cubozoan). Although superficially resembling jellyfish, they are not closely related to them. They get their name from their box-like appearance and four tentacles. It's difficult to determine the deadliest animals in the world; there are many capable of quickly killing someone, but these animals would certainly be near the top of anyone's list. The venom from a single sea wasp (*Chironex fleckeri*) could kill up to 60 adults within 4 minutes! Luckily for us, these are not aggressive animals. Their toxin has evolved to kill their prey, which includes small fish, as quickly as possible to prevent them from being damaged. Sea Wasp are found in the Indo-Pacific regions and fortunately fatal encounters are rare. While most Cnidarians only possess very primitive "eyes", which are little more than light-sensitive cells, sea wasps have more sophisticated eyes capable of forming an image to aid in hunting down their prey.

Sea anemones are in the class Anthozoa, which literally means 'flower animals'. They are well known to the public, especially since the movie *Finding Nemo* correctly portrayed clownfish (*Amphiprion*) as having a symbiotic relationship with sea anemones. There are about 1,000 species of anemones, and like all other Cnidarians, sea anemones possess stinging tentacles used to capture their small prey. Clownfish have evolved mutualistic symbiotic relationships with about 10 species of sea anemones in the Indo-Pacific regions. To prevent them from being stung, clownfish produce a thick layer of mucus. In this symbiotic relationship, both species benefit; the anemone protects the clownfish from predators and the clownfish cleans the anemone of waste and potential parasites.

Corals are close relatives to sea anemones and also deserve the name "flower animals". Similar to flowering plants they come in a variety of shapes and brilliant colors that would rival any terrestrial ecosystem. Their bright colors are caused by pigments produced by the coral to protect them from the harmful effects of UV light. Worldwide, there are more 5100 described species of corals. Although, the true extent of their diversity has yet to be realized because many could be cryptic species. Cryptic species closely resemble each other in appearance due to convergent evolution, but may not be closely related. Recent molecular work has indicated there could be many more species.

Corals are among the most important groups of organisms in the oceans for their role in creating diverse ecosystems. They are typically small animals forming large colonies by extracting calcium carbonate from the water and then excreting it as rock. Overtime, the rocks grow in size eventually forming large coral reefs in the tropical regions of the world. While coral reefs may account for less than 1% of the ocean's area, it has been estimated that nearly half the ocean's diversity may be dependent on coral reefs. Similar to trees providing additional habitat on land, coral reefs provide numerous types of habitat in regions that would otherwise be a biological desert.

There's a paradox of coral reefs; they are among the most diverse and productive ecosystems in the world, yet they exist only in crystal clear water. The paradox lies in the fact that clear water lacks nutrients that many organisms need to grow, including phytoplankton. Phytoplankton, being autotrophic, often forms the base of most marine ecosystems supplying nutrients and energy to all the other animals. However, most species of corals require lots of sunlight to grow, and nutrient rich waters would promote algae and phytoplankton growth, which would reduce light penetration or would compete directly with the corals. Despite the lack of nutrients in the water, coral reefs thrive in the nutrient poor conditions, rivaling the tropical rain forests in diversity and ecological complexity.

The success of corals in nutrient poor waters is dependent on their symbiotic relationship with tiny photosynthetic eukaryotes called zooxanthellae that live inside the coral's cells at densities approaching a million cells per square inch. Here's another example of a symbiotic relationship in nature when both organisms benefit from the arrangement of two organisms living together. The corals send out their small tentacles collecting tiny crustaceans and plankton floating in the water, and the zooxanthellae produces carbohydrates to supply the corals. If there are too many nutrients in the water, then algae may overtake the corals killing them. In recent years, coral reefs have suffered major losses due to a variety of sources, which I will discuss further in chapter 14 Conservation Biology.

Mollusks
From the Primitive to the Advanced

Mollusks are an incredibly varied group of animals with over 100,000 species currently known. They include the chitons, snails, bivalves, and squids and octopi (cephalopods). Most mollusks have a shell made of calcium carbonate, similar to the reef-building corals. However, several groups of mollusks have independently lost their shells. The cephalopods provide a great example illustrating how certain groups have evolved reduced shells. The chambered nautilus, the most ancient group of cephalopods has retained their shells, in squid their shells are reduced to a stiff pin, and octopus have totally lost all trace of a shell. Some groups of snails (gastropods) have also lost their shells. This includes the very brightly colored nudibranchs (sea slugs) and the common garden slug.

Even with the incredible diversity of mollusks, there are several features they all have in common, inherited from their last common ancestor over 500 million years ago. The most obvious feature is a large body cavity called the mantle that they can also use for breathing. All mollusks have a unique mouth part called a radula, which has been modified in each group to accomodate their feeding habits. In most snails, the radula is used to scrape algae from rocks and other hard surfaces. It is modified into a horny beak made of keratin in cephalopods and can be used to eat fish and crabs.

Most people are quite familiar with snails, a unique group of animals that produce a a protective shell made of calcium carbonate. In marine environments, snail shells can be quite large, colorful, and covered in unique patterns. An interesting aspect of snail shells is that their shape and color patterns follow precise mathematical sequences such as the Golden Mean or Fibonacci sequences. Snails have varied lifestyles, many eat detritus, aquatic plants, and algae, while others are predacious. Most snails are harmless except for the cone snail (*Conus*), which is one of the deadliest animals on the planet. A tiny amount of the neurotoxic venom is capable of quickly killing an adult. Scientists are currently studying their venom to learn how our own nerve cells work.

Not all snails have shells, including the common garden slug and the often highly colorful marine nudibranchs (nudi = naked, branch = gills or naked gills), also known as sea slugs. The bright colors of nudibranchs are a form of aposematic warning coloration, which serves as a warning to other animals not to eat them. Many are poisonous or taste very bad, so predators leave them alone (same as in monarch butterflies and coral snakes).

A relative of the sea slug, with the name *Elysia chlorotica*, eats algae at a very early age and then does something quite remarkable; it doesn't fully digest the algae. Instead it incorporates the chloroplasts into its own cells and takes on a green appearance. The result is that this animal is capable of photosynthesis. It can go a long time without feeding when in lots of sunlight. Additionally, its cells have actually evolved the ability to support the chloroplasts keeping them photosynthesizing for months.

The cephalopods are an interesting class of mollusks that include the chambered nautilus, squid, cuttlefish, and the octopus. By several standards, cephalopods are highly advanced. With the exception of the chambered nautilus, most have lost their shells over time. By losing their shells, they have lost protection, but they have gained speed and maneuverability. Cephalopods include the largest of all invertebrates, the colossal squid (*Mesonychoteuthis hamiltoni*), found in the southern oceans. It may reach a length up to 46 feet and weigh up to 1600 pounds. Very little is known about its life history and other habits. It is believed to be an ambush predator of deep sea fish, in turn, it is prey for sperm whales. The cephalopod eye evolved independently of the vertebrate eye, but operates fundamentally in the same way and is capable of the same high level of vision as vertebrates.

The octopus has been at the center of a controversy over its perceived intelligence. Judging and measuring intelligence in other animals is a difficult task confounded by the fact that there is no universal test of intelligence and many researches disagree over what factors are important for determining intelligence. Despite the difficulties in defining and studying animal intelligence, the octopus is an active predator, often eating crabs. Some octopus have been known to locate a crab and wait to ambush it when it emerges from its hiding place, indicating forethought. Interestingly, some cephalopods actively change their skin color for communicating emotions, confusing its prey, and of course for camouflage. Squid have been observed to participate in cooperative hunting for fish. To top it off, cephalopods have achieved these abilities with a very different nervous system and brain structure compared to vertebrates.

The Arthropods: a Body Plan for Success

It's difficult to determine which group of animals is the most successful, it may simply depend on your criteria. If you were to choose success based on diversity and sheer numbers of individuals, then arthropods could be considered the most successful group of animals on the planet. To a first approximation, the average animal is an arthropod. Currently, there are more than a million described species of arthropods with new species constantly being discovered. Perhaps the main factor accounting for their success is their hard exoskeleton with jointed limbs. It doesn't take much change in an arthropod's DNA, only a few small mutations can lead to changes in their body, quickly leading to a new species. When combined with rapid generation times, arthropods are capable of quickly evolving to adapt to a changing environment or to exploit a new resource.

The arthropod exoskeleton is made of a tough substance called chitin, the same chemical that forms the cell wall of fungi and is a chemical relative of cellulose, which forms the cell wall of plants. All arthropods routinely shed their exoskeleton, allowing them to grow each time they molt. Despite the success of arthropods, the exoskeleton, places a size limitation on arthropods. Luckily, we will never have to face a giant 50-foot praying mantis, the result of mutations from radioactive fallout as portrayed in 1950s science fiction films. As an animal gets larger, so must its support, but the scaling is not linear. Meaning, if I were to take a mouse and enlarge it proportionally to the size of an elephant, its bones would instantly break from its own weight. If arthropods were to grow much larger, their exoskeletons would have to rapidly increase in thickness. At some point, the exoskeleton would be so large, that the animal would be unable to move.

Despite the size limitations on arthropods, some individual species have become relatively large. The first arthropods colonized the land about 425 million years ago, and by 400 million years ago, giant centipedes 6 feet long roamed the land. In the ocean, giant sea scorpions reaching 7 feet long swam in warm coastal waters preying on fish and marine arthropods. These ancient arthropods were able grow larger, potentially reaching the upward limits of their size, because there was more oxygen in the atmosphere. Arthropods obtain oxygen through simple diffusion; if the animal grows too large, then it won't get enough oxygen to all its internal organs and muscles. But, 400 million years, there was about 8% more oxygen in the atmosphere, enough to let those ancient arthropods grow to about their maximum size possible.

Arthropods are well known for their compound eyes made up of tens to thousands of individual lenses called an ommatidia. Their eyes work much like pixels in a digital camera, resolution improves with more pixels. In arthropods, the more ommatidia means the better their vision. Contrary to what's portrayed in the movies, arthropods do not see numerous multiple images, it's more like a low resolution photograph.

There are four major groups of living arthropods, the myriopoda (centipedes and millipedes), the chelicerates (horseshoe crabs, and arachnids), the crustaceans (shrimp, crabs, and many others), and the insects, the most species-rich group on the planet. Trilobites represent a fifth extinct group of arthropods that existed for nearly 300 million years before their extinction 251 million years ago.

Chelicerates
Horseshoe Crabs and Arachnids

You may not have heard the term chelicerate before, but you are likely very familiar with them as they include spiders, ticks, scorpions, and the prehistoric looking horseshoe crabs. The name of this group is based on the presence of chelicerae, which are appendages found in front of the mouth and shared by all members of this group. As you may have guessed, they inherited this feature from a distant common ancestor that lived around 500 million years ago.

In most chelicerates the chelicerae are simple pincers used for feeding. In spiders they are modified into fangs used to inject venom into their prey. All Chelicerates, including Arachnids lack antennae, easily distinguishing them from insects and crustaceans. They have two main body parts, the abdomen and a cephalothorax. As its name indicates, the cephalothorax is formed from the fusion of the head (cephalo) with the thorax (head – abdomen). There are three major groups of chelicerates; the horseshoe crabs, sea spiders, and arachnids which is the most diverse lineage including spiders, scorpions, wind scorpions, and ticks.

Perhaps one of the oldest living animals are the horseshoe crabs (*Limulus polyphemus*). They are not related to true crabs, but are more closely related to spiders based on the presence of chelicerae and the cephalothorax. Fossils of ancient horseshoe crabs show that they have changed very little in 450 million years, they really are a living fossil. You can find horseshoe crabs on the Atlantic and Gulf coast shores where they emerge by the thousands from the surf during spring high tides to lay their eggs. Each year, as they have done for hundreds of millions of years, horseshoe crabs will lay their eggs all at once right at the top of the high water line. Shorebirds have timed their migrations to coincide with the egg laying to take advantage of the nutritious eggs. In recent years, horseshoe crab numbers have dwindled solely due to overharvesting. The loss of horseshoe crabs and their eggs has resulted in significant losses of shorebird populations, especially the Red Knott.

Most arachnids live on land and are among most feared animals on the planet. It's been estimated that 52% of all women and 35% of all men suffer arachnophobia, the fear of spiders. This is unfortunate because very few spiders are capable of harming humans, and spider bites from those species are quite rare. Spiders are a vital part of most terrestrial ecosystems because they are voracious insect predators. Although, some spiders are large enough to prey on small rodents, birds, and one spider called the fishing spider lives at the edge of water and preys on small fish.

As you already know, many species of spiders build intricate webs to catch their prey. They use silk made of proteins, which is stronger than steel, to spin their webs. Because the webs are sticky and collect dust, reducing their ability to trap insects, the spiders will eat their own web to recycle the proteins, only to continually rebuild them. Not all spiders build webs to catch prey, small jumping spiders in the family Salticidae, which literally translates to "jumping", are active predators constantly on the hunt for prey. They have two large forward facing ocelli, which function as eyes giving them great depth perception, similar to humans. In the late fall, when outside temperatures begin to drop, many spiders will move indoors. The small jumping spiders are actually beneficial because they don't carry diseases, don't build webs, and will catch numerous unwanted small insects in your kitchen and home.

Other arachnids include the scorpions and wind scorpions. Scorpions are uniquely shaped predators with a long and narrow abdomen that has been greatly modified to have a stinger at its tip. They are capable of delivering a powerful sting, which it uses to paralyze its prey. Wind scorpions are among the fastest of all arachnids allowing them to chase down their prey. Some species in Middle Eastern deserts, known as camel spiders, can reach 6-inches. Wind scorpions have the unique distinction of possessing the largest jaws compared to its body of any animal. While not venomous, their jaws are over twice the size of their head allowing them to inflict quite a powerful bite as they shred apart their prey. Ticks and mites are also arachnids, they are ectoparasites, feeding off the blood of other animals. Unfortunately, some ticks can carry diseases that are harmful to humans.

Crustaceans
Include the Smallest and the Largest Arthropods

Like the title states, crustaceans include the largest and smallest of all living arthropods. The vast diversity of crustaceans is found in the oceans, but a few are found in freshwater lakes and rivers, and the familiar pill bugs are commonly found on land throughout much of North America. Crustaceans are somewhat similar to insects because both groups have antennae, although crustaceans often have two pairs. Crustaceans have more legs than insects along with appendages on their abdomen, and crustaceans never developed flight. Crustaceans are an incredibly diverse lineage with over 67,000 known species. They include the tiny copepods, sessile barnacles, primitive looking tadpole and fairy shrimp, and the more familiar crabs, shrimp, and lobsters.

Many crustaceans are very small. Among the smallest of all animals are the copepods found in freshwater and marine ecosystems world-wide. Other small crustaceans live in ephemeral ponds, which are temporary water bodies only present part of the year. In the desert southwest during the summer monsoon season, these small temporary ponds fill with water and quickly become populated with tadpole shrimp (*Triops*), another living fossil whose form has not changed in 300 million years. In a matter of weeks, they will hatch from eggs, grow, mate, and lay their eggs, quickly completing their entire life cycle before the ponds become dry. When the ponds dry up, strong winds can blow their eggs for many miles spreading the species over large geographical areas. These ephemeral ponds are also home to small fairy shrimp, a diverse lineage of small crustaceans that appear to swim upside down. Tadpole shrimp and fairy shrimp are similar in appearance to some of the earliest crustaceans dating back to the Cambrian. Perhaps, the reason they still survive today is by living in ponds without fish that would quickly eat them.

Perhaps the best known crustaceans belong to the Decapods (deca=ten, pod=foot), which includes shrimp, crabs, and lobster. Based on this list, it becomes apparent why they are well known, for some people, they are quite tasty. The largest living arthropod is the giant Japanese spider crab (*Macrocheira kaempferi*) with a leg span measuring almost 18 feet across. Crabs are quite diverse and are mostly marine. Hermit crabs carry snail shells on their backs for protection. Sometimes they will acquire small sea anemones to help camouflage them from predators. Small hermit crabs often go about trading each other's shells several times a day. Often seen running on sandy beaches around the world are the white looking ghost crabs, quickly darting into their holes in the sand if approached too quickly.

Shrimp also come in many sizes shapes and colors. Small krill, less than an inch in length serve as the main food for whales, the largest animals on the planet. On tropical reefs is a red and white striped shrimp with two large pairs of white antennae, these are the cleaner shrimp (*Lysmata amboinensis*). At cleaning stations, fish will line up allowing the cleaner shrimp to enter their mouth and gills removing any visible parasites from the fish. The relationship is mutually beneficial for both the shrimp and the fish, the fish gets pesky parasites removed and the shrimp gets a meal. Another small shrimp called a pistol shrimp has a unique claw capable of producing a loud snapping sound. They form a mutualistic relationship with small fish called shrimp gobies. The pistol shrimp digs a hole where they both live and the goby, with its good eyesight, protects the pistol shrimp.

Considered a delicacy by many, lobsters are also a decapod. In North America, there are several species; the American lobsters (*Homarus americaus*) have two large claws and are one of the more commonly encountered commercial species. In Florida and the Caribbiean is the spiny lobster (*Panulis argus*), which lacks the large claws. Spiny Lobsters are known to go on large migrations to their spawning grounds. They will follow each other in single file lines, maintaining contact through their antennas. An interesting fact about lobsters is that they can live up to 70 years. Their life span is actually more a function of growing too large, so when very large lobsters molt, the energy expenditure can actually kill them through exhaustion. Another interesting fact about lobsters is that they don't really grow old and senesce similar to humans. The reason is because they express telomerase as adults, which means that there is no limit to the number of times their cells can reproduce.

Related to decapods are about 400 species of the colorful mantis shrimps. These voracious predators have two unique adaptations. First, their large eyes are among the most complex visual systems in the animal kingdom with 16 different photoreceptor pigments, compared to three photoreceptors in our eyes. Second, they have incredibly powerful raptor-like claws that move so fast that it sends a shock wave through the water with enough power that it stuns or even kills their prey. Some mantis shrimp are able to attack animals with hard shells and there are reports that they can break the glass of home aquariums.

Nature has an Inordinate
Fondness for Insects

Insects are found on every continent and are by far the most diverse lineage of animals on the planet with over one million species currently described. They originally evolved on land over 400 million years ago during the Devonian, before vertebrates colonized land. Insects have continually been diversifying ever since. Flight evolved one time in all the invertebrates, and it was the insects that did it. With their diversity has come many novel adaptions in the animal world. Ants, bees and wasp have evolved to form some of the most complex societies on the planet. Butterflies and beetles undergo complex metamorphosis so that the larvae and the adults look and behave very differently from each other. And some insects have aquatic larvae and the adults spend their life on land. For the remaining part of this chapter, I cover some of the fascinating features of a few insects, just enough to provide a tiny glimpse into their diversity. At the end of the chapter, I will also reveal the most dangerous insect in the world.

Dragonflies may well be one the best predators in the world based on an incredibly efficient catch rate of 90-95%, far better than other well-known predators such as cheetahs (about 50% success rate). Another interesting feature of dragonflies is that they have changed very little since their first appearance in the fossil record about 325 million years ago, it seems as if nature got their design right from the start. Today, there are about 3,000 species of dragonflies. They have large eyes for an insect, allowing them to see small prey on the wing. Their legs are well adapted to catching insects in the air, but they are incapable of walking once they have landed. Dragonflies are also some of the best flying animals in the world, they have two pairs of wings capable of independent movement. In fact, dragonflies are so good at flying, the U.S. military is currently studying their wings to develop new flying technology. Larval dragonflies are also adept hunters as well. Unlike the adults, larval dragonflies are aquatic, confined to hunting prey underwater. Larval dragonflies have a unique mouth part called the prementum that shoots forward to catch their prey.

Stoneflies, mayflies, caddisflies are three groups of insects, in addition to dragonflies, with winged adults and the larvae are confined to aquatic habitats. Mayflies belong to the order Ephemeroptera, which means fleeting (ephemera) wing (ptera). The larval stage of mayflies may last for several months to several years. Then, thousands of the larvae will simultaneously crawl out of the water, and molt into adults by the tens of thousands. The adults do not feed, and must find a mate before they die, all within a day. By timing their emergence all at once, they can overwhelm their predators and increase their chances of mating.

Caddisflies are known for making casings out of small pebbles, leaves, or debris from their environment. They are related to the butterflies and moths, which is evident by the fact that they produce silk to form their casings and the adults are easily mistaken for small moths. Stonefly larvae are commonly found in clean running waters where they are a favorite food of trout. Scientists often use these three groups as indicators of stream quality because most of them are adversely affected when water quality declines.

There is a group of insects actually called the true bugs. They are very diverse, but all of them have a mouth part called a rostrum that is used for piercing and sucking. Aphids, small true bugs found on many garden plants, insert their rostrum into the plant's phloem. The phloem is part of a plants vascular system used to transport sugars from the leaves to the rest to the roots. Aphids can be seen lined up on plants with their heads down, constantly drinking from the plants. If you look closely, you may notice ants tending and protecting the aphids. The aphids supply the ants with sugar water from the plant, in turn, the ants protect the aphids, another example of symbiosis. Assassin bugs hunt insects on flowers and uses its rostrum to suck fluids out of their prey.

During summer nights, the forests come alive with sound of cicadas. They actually spend most of their lives as nymphs underground, one species even spends 17 years in this stage. Eventually, the nymphs will crawl up on a tree and molt into adults. Their song is produced from modifications in their abdomen, which is mostly empty space.

Butterflies and moths form one of the most recognizable groups of insects in the word. In North America alone, there are over 800 species of butterflies and nearly 10,000 known species of moths. These insects undergo a process called complete metamorphosis when the larvae restructure their body as they transform into an adult. Butterflies and moths both form a cocoon to protect themselves during this vulnerable time. The reason for this radical change in body form is an evolutionary strategy to prevent the adults from competing with the larvae over the same resources. For example, butterfly caterpillars feed on leaves, whereas the adults feed mostly on nectar.

Many species of moths and butterflies will lay their eggs on only certain species of plants that they have been coevolving with for millions of years. Unable to move rapidly, caterpillars have evolved defenses to protect themselves from predators and parasitoids; including camouflage, bitter taste, poisonous hairs and various projections capable of inflecting an annoying sting or irritation. Butterflies and moths are also a prime example of adaptive radiation, which is an increase in diversity that has coincided with the evolution of flowering plants.

Often feared by humans for their powerful stings, ants, bees, and wasps are a fascinating group that includes some of the most complex societies on the planet and unique mating systems. Ants may be one of the most important insects in the world. They are known to form large colonies, sometimes exceeding a million or more individuals. Their burrows can aerate the soil and they are important for dispersing seeds. About 200 species of army ants found in the tropics form mobile colonies carrying all their eggs and the queen as they move through the jungle. They will often quickly invade an area, known as a raid, where they will forage and eat almost any small animal in their path that are unable to get away from them.

With their division of labor and large size, leaf cutter ants form some of the most complex societies outside of humans. They evolved the ability to farm fungi long before humans. Large workers can be seen constantly bringing leaf parts into the colony, which are used to cultivate and grow a particular fungus eaten by the ants. Small workers, can be seen riding on top of the leaves carried by the much larger workers, their function is to protect the larger workers from tiny parasitoid flies and wasps. Other workers remove waste from the colony while others tend the fungus gardens.

Bees and wasp are feared for their powerful stings from a modified body part called an ovipositor, which is normally used to lay eggs. Not all wasps form large colonies, some are solitary and use their ovipositor to lay their eggs in other animals or plants. They are called parasitoids, which is slightly different than a parasite. Parasites, typically don't kill their host. But, a parasitoid will. In the desert southwest are the tarantula hawk wasps, a group of very large solitary wasps that hunts tarantulas to lay their eggs. Upon finding a tarantula, a tarantula hawk will sting the spider, not to kill it, but to paralyze it. Once it's paralyzed by her stings, she will then drag the spider down its own hole, lay her eggs in it, and then bury the spider, alive. The eggs will hatch and the larval wasp will eat the spider from the inside out, only to kill it once the wasp emerges as an adult.

Beetles may be the most diverse order of insects on the planet, it has been estimated that one out of every four animals on this planet is a beetle! Similar to butterflies and moths, their increase in diversity over the past 100 million years is also tied to the evolution of flowering plants. Beetles live everywhere, and there are many species of beetles yet to be described by scientists. In a famous experiment to determine insect diversity, a scientist fumigated a single tree in the tropics to estimate how many insects were in the canopy. Upon collecting all the insects, there were about 50 species of beetles that had never been collected or observed by scientists. Beetles can be both beneficial and harmful to humans, some species of beetles are crop pest, while others like the lady bugs are predators of crop pests. Other beetles are important pollinators.

Flies are another very diverse group of insects with most species being completely harmless and many are actually beneficial. Unlike most other insects, flies have a single pair of wings and the second pair is reduced to small halteres that function much like a gyroscope. Fortunately, not all flies are human pests. The small fruit fly has been used for over a hundred years as a model organism to study genetics. Flower flies are harmless and are important pollinators in many ecosystems, some have even evolved to mimic local bumblebees so that predators will leave them alone. However, robber flies, which also mimic bees, are ambush predators of insects.

What is the most dangerous insect on the planet? If you ask someone who studies diseases, they will tell you it is mosquitoes, a type of fly. It's not that mosquitoes are super venomous or produce a super toxic poison. Instead, their danger lies in the fact that mosquitoes and other flies transmit some of the deadliest human diseases, including malaria, west Nile virus, dengue fever, and yellow fever. It has been estimated that each year 700 million people get some disease from mosquitos, killing over one million people.

Echinoderms
Relatives to Chordates

It may be hard to believe, but starfish and sea urchins, seemingly simple animals in the phylum Echinodermata, are the closest living relatives to chordates, the phylum which includes us. The close relationship was originally based on similarities in early growth and development, and later further supported by DNA evidence. Echinoderms are a varied lineage with familiar animals including, sea urchins, sea cucumbers, starfish, and sand dollars. Their radial symmetry is only superficial, in fact the ancestors to star fish were actually bilateral. I often wonder what caused these animals to forsake bilateral symmetry and evolve radial symmetry.

There are about 7,000 species of Echinoderms, which are only found in marine environments and often in great numbers. Similar to cnidarians, which also have radial symmetry, echinoderms lack eyes, a head, and centralized sense organs. However, the tube-feet of starfish and sea urchins contain a light-sensitive protein called opsin, allowing them to detect areas of dark and light. Mammals also contain opsin in their eyes as well, which is used for the same purpose. Starfish can be seen slowly moving with one of their arms slightly lifted in the direction they are moving.

Sea urchins are grazers of marine algae and can reach very large numbers. In the pacific coast of California, sea urchin populations are kept in check by sea otters, their biggest predators. Sea otters live among kelp forest where they dive down in search of urchins. Once they find one, the otter will return to the surface, float on its back to crack it open and eat the tasty insides. By eating the sea urchins, otters make it possible for the large kelp forests to exists, thus greatly increasing diversity of the nearshore marine ecosystems along the west coast of North America, from California to Alaska.

Sea cucumbers are less well known than sea urchins and starfish, but they have a very unique behavior when threatened with predation. When attacked, they will eviscerate themselves and then regrow their insides! Starfish are well known echinoderms, which can live up to thirty-five years. Some reproduce asexually by physically detaching one of their five arms. The starfish will then regrow the arm, but even more remarkably, the severed arm will also grow into a new starfish! Another fun fact about a starfish is that some will eat with their stomach inside out. When a starfish approaches a bivalve, it will use its arms to pry the shells open, then forces its stomach into the prey to digest it. This way the starfish can eat larger prey without having to actually ingest the whole animal.

The Worm-like Animals

There are many other major groups of animals that I have not covered in this chapter. Many are wormlike, lacking appendages. Despite their vermiform shape, they are not necessarily closely related to each other. Unfortunately, it is impossible to cover their diversity here. But, I have included a brief description of the larger groups.

The Flatworms worms have a flat appearance, which increases their surface area compared to their volume. It's an important adaptation considering most of these worms are parasites including liver flukes, blood flukes, and tapeworms. By having a large surface area, a parasitic flatworm can absorb all the nutrients in needs across its skin. Tapeworms live inside the gut of vertebrates. If you eat raw or undercooked meat, you could become infected with a tapeworm that can grow up to 50 feet long and live for 30 years. During that time, they will produce millions of eggs that will be released whenever you go to the bathroom. Blood flukes cause schistosomiasis, which is a major problem in the tropical regions of the world.

Horsehair worms are parasitic worms that resemble the coarse hair of a horse's mane. Any insect infected by a horsehair worm is doomed to die. Unsuspecting insects will eat the eggs of the worm. Once ingested, the eggs will hatch and the worm will grow inside the abdomen until it reaches maturity. At this time, it will cause the insect to jump into water and drown itself and the worm will burst out of the insect's abdomen. It will then find a mate and lay their eggs along the edge of the puddle, waiting for the next insect to continue the cycle.

Annelids are the segmented worms and include the familiar earth worm found in gardens throughout North America. They are also a favorite of freshwater fishermen and are found at any local bait and tackle store. Earthworms are basically a muscular tube that burrows through the ground ingesting organic material. They are important for aerating the soil. Leeches are relatives of earth worms, although they are an ectoparasites, but not really a harmful once. Fortunately, they are not known to spread diseases. In fact, they have been studied for potential medical benefits.

Nematodes are the round worms with at least 25,000 species, half of which are parasitic, including heart worms. They are found in almost every ecosystem and in great numbers. It has been estimated that they may account for 8-10% of all the animals on Earth.

Chapter 10

Vertebrate Diversity

Bigger, Faster, Smarter
The vertebrates have done well

Introduction

It can be difficult to choose the most successful lineage of animals on the planet. There are about 64,000 species of vertebrates, which includes the largest animal on the planet, the fastest animals on the land, in the air, and in the water. And of course, it includes humans, clearly the most intelligent organism to have ever inhabited the Earth.

Modern vertebrate diversity is the result of more than 500 million years of evolution, extinction, and geological forces. Vertebrates have successfully colonized every continent and are found in every ocean. They range in size from a tiny minnow less than a third of an inch in length, to the blue whale, reaching lengths over 100 feet. Although vertebrates got their start in marine environments, one lineage, the tetrapods have successfully colonized the land, and have taken to the air at leas three separate times. What's interesting is that tetrapod's have routinely been able to successfully return to the oceans, as seen by marine mammals and sea turtles, along with several groups of extinct reptiles from the Mesozoic Era. Vertebrates also evolved to be warm blooded in birds, mammals, and even in some fish. By generating body heat internally, they can be active in any environment, regardless of the temperature.

Since the first simple chordate, the evolution of an internal skeleton, hard teeth, jaws, paired appendages, swim bladders, and lungs helps to explain their success as these features have been continuously modified and repurposed over time. For example, fins evolved into limbs, scales into feathers and hair. Swim bladders used by fish to regulate buoyancy evolved into lungs for breathing on land. Jaws used to improve oxygen absorption in water also evolved sharp teeth to catch more prey. Two-chambered hearts evolved into three chambers, then four chambers, maximizing oxygen delivery to increasingly metabolically demanding muscles and larger animals. An internal skeleton providing muscle attachment for rapid swimming was strengthened to live on land. And so it goes, the vertebrates continually evolved and diversified into an incredibly successful group.

A Brief History of Vertebrate Evolution

The vertebrate story possibly began with a small wormlike animal called *Pikaia* swimming in the ancient Cambrian oceans over 500 million years ago. Despite its simple structure, *Pikaia* had many of the features still found in vertebrates today that unite this entire phylum to a single common ancestor. Based on fossil evidence, molecular evidence, and shared characteristics, it is clear that all modern vertebrates descended from a single common ancestor.

All vertebrates possess pharyngeal slits, a dorsal nerve cord, a notochord, a muscular post-anal tail, and mineralized tissues including teeth, cartilage, and bones. Although, I should point out that some of these features have become highly modified. But most importantly, vertebrates possess a vertebral column made of vertebrae to protect the nerve cord and provide rigid support for muscle attachment. In primitive chordates, they lacked a vertebrae and the notochord was a stiff cartilaginous rod running parallel to the nerve cord. The main purpose of the notochord was to support the body and provide a structure for muscle attachment helping it swim by undulating its tail. By having a stiff rod running the length of the animal, its body would not shrink when the muscles contracted, allowing it to quickly swim through the water.

Starting about 450 million years ago in the Ordovician a wave of major evolutionary innovations began to take place with the origins of jaws, teeth, and paired appendages (pectoral and pelvic fins). Natural selection is similar to "tinkering". Over time, the ancestral features of vertebrates became increasingly modified, as vertebrates have continuously evolved to exploit new resources in a constantly changing world. A great example of evolutionary tinkering has occurred in the pharyngeal (gill) slits found in the sessile tunicate where they are used for filter feeding. In the earliest fish, the pharyngeal slits were first modified into gills to improve gas exchange, getting more oxygen to metabolically demanding muscles. With the evolution of gills for gas exchange, our ancestors went from stationary filter feeders to active predators. The front gill slites later evolved into jaws that at first may have been primarily an adaptation for even greater water flow over the gills, allowing for even more physical activity due to greater oxygen flow. But, sometimes serendipity happens in evolution; the same jaws that may have evolved for improved gas exchange in the water also allowed fish to bite harder, making them capable of taking a larger variety of prey.

These first fish with jaws are called Gnathostomes, which literally means "jawed mouth". They had sharp teeth and when combined with paired fins, they were formidable predators for their time. In fact, the rise of jawed fish was probably responsible for the demise and extinction of the more primitive Ostracoderms and other jawless fish that once dominated the ancient Devonian oceans. The fact that Ostracoderms lacked jaws but had armor made of teeth-like structures, indicate that they may have been a transitional lineage, filling the transition between jawless fish and jawed fish. Teeth and other mineralized tissues, such as bones and cartilage, are a mineral you probably never heard of called hydroxyapatite. It is comprised of calcium and phosphorous and is much tougher than calcium carbonate, which is used my mollusks and reef-building corals.

During the Ordovician, Jawed fish had split into two lineages, the cartilaginous fish and the bony fish or Osteichthyes. The cartilaginous fish include the sharks, skates, and rays. Modern sharks have changed very little from ancestors going back 360 million years. When it comes to diversity, the bony fish are the clear winners. Their success is likely tied to the evolution of a gas-filled swim bladder allowing the fish to control its buoyancy in water. A swim bladder saved the fish energy from constantly having to swim and freed the fins allowing greater maneuverability.

Shortly after their arrival, bony fish split into two major lineages, the lobe-finned fish and the ray-finned fish. Today, when we think of fish, it's the ray-finned fish that come to mind. There are about 30,000 species of ray-finned fish inhabiting every part of the oceans from crushing depths of more than 10 miles below the surface, to the near freezing waters of the polar regions. Not to mention fish are also found in freshwater lakes and rivers throughout the world.

By the time of the Devonian, 419 to 359 million years ago, fish had become quite diverse, with some scientists calling it the "Age of Fishes". About 375 million years ago, a particular species of lobe-finned fishes evolved into the first tetrapods (tetra=4, pods=foot), and the colonization of the land by vertebrates had begun. Once again serendipity and evolutionary tinkering become apparent as old features are modified to function in new environments. The swim bladder had evolved into lungs and was further modified for a life on land. The pectoral and pelvic fins were modified into the four limbs of modern tetrapods. Even the ancient bone arrangement of one bone, two bones, and lots of bones, has been preserved in all the major lineages of tetrapods. The pharyngeal gill slits were no longer needed for gills and evolved into parts of the jaws, necks, and even ears in mammals. The bones that mineralized bones used for muscle attachment in fish became larger and strengthened to support vertebrates on land. Over time, tetrapods greatly diversified into modern amphibians, reptiles, birds, and mammals. What that means, is that we are descended from a fish ancestor. This may seem like a far-fetched claim, but the evidence is in recorded in the fossil record, seen in our own anatomy, and written in our DNA.

Sharks, Skates, and Rays
How Nature got it Right

The first cartilaginous fish have inhabited the Earth over four hundred million years ago. Today, there are close to 900 living species, which mostly include the sharks and rays. The first sharks appeared about 420 million years ago and have changed very little since, it's as if nature got sharks right from the start. Although, the first truly modern sharks date back to the Jurassic about 144 million years ago. The major difference being in the position of the mouth; modern sharks possess a much more pronounced rostrum, possibly an adaptation for small electro-receptors used to detect their prey. The ancestors of rays, easily identified by their flattened body, diverged from the sharks about 200 million years ago.

Despite their lack of diversity compared to the bony fish, sharks and rays are an incredibly successful group. First, they represent one of the earliest radiations of jawed fish over 420 million years ago. Then, about 360 million years ago, they underwent another round of rapid radiation in diversity, which may have led to the demise of the once successful placoderms. The placoderms were an ancient relative of sharks with hard armored coverings that once dominated the Devonian oceans during the Age of Fishes. Today, cartilaginous fish are still successful, they are among the most common apex predator in the world's oceans. Although over-fishing has drastically reduced their populations, making many species vulnerable to extinction. They reproduce slowly and the Greenland shark (*Somniosus microcephalus*) may live to 400 years old, making it the oldest vertebrate animal.

Cartilaginous fish get their name from the lack calcified or mineralized bones, so complete fossils are rare. Surprisingly, there has been some recent DNA evidence indicating that their lack of hard mineralized bones may actually be a derived trait. The implication is that their ancestors forsake mineralized bone, which may have provided them greater agility. However, their teeth are made of the hard mineral hydroxyapatite, the same mineral used by all jawed fish, including ourselves, to make teeth. Shark teeth are abundant in the fossil record and are sometimes the only evidence we have of an extinct species.

It's well known that some species of sharks have powerful jaws filled with lots of teeth for catching their prey. The size and shape of their teeth are related to their size and the type of prey they consume. Shark teeth are constantly replaced and a single shark can shed upwards of 20,000 teeth in a lifetime, another reason why shark teeth are common in the fossil record. What you may not know is that the scales of sharks and all cartilaginous fish are actually miniature versions of teeth called placoid scales. It takes only small changes in a shark's DNA to change the shape of a tooth into a smaller placoid scale. It's just a matter of changing its shape and size, everything else is the same when it comes to the structure of a shark tooth. In fact, their teeth and scales have dentine and enamel, just like our teeth.

Cartilaginous fish lack a swim bladder, therefore they must constantly swim, or they will sink to the bottom of the ocean. You may have also noticed that many sharks seem to swim menacingly with their mouths open revealing their rows of sharp teeth. They swim like this because, similar to all other fish, they must constantly move water across their gills for gas exchange. When sharks swim with their mouth open, it forces water across their gills in what is known as ram ventilation. If a shark becomes tangled in a net or caught on a long-line, then it can drown because it is unable to actively move water across its gills without swimming. Their pectoral fins are enlarged and provide additional lift while they swim. Some sharks, including nurse sharks, and almost all rays have evolved the ability to actively move water across their gills by pumping water through their mouths and through small openings near the eyes called spiracles.

Very little is still known about shark reproduction in terms of where they reproduce, how often they reproduce, and how potential mates find each other. We do know that sharks and rays either lay eggs while others give live birth to small pups. Live birth in sharks is different from mammals. In sharks that give live birth, the eggs remain inside the female until they hatch. Sometimes, the baby sharks, may remain inside the mother growing larger to improve their chances of survival when they go off on their own. There is evidence that the pups may actually eat each other with only the largest individuals surviving to leave the mom. Unlike most ray-finned fish, sharks reproduce internally, a derived trait compared to their ancestors.

With the advent of GPS, scientists have attached satellite trackers to sharks which have revealed new information about their movements and habits in the last few years. Some of the interesting discoveries relate to their world-wide movements as they swim from the coasts of California to Hawaii, or from South Africa to Australia, sometime swimming in total darkness more than 2500 feet below the surface for thousands of miles. No one is quite sure why they make such long migrations or how they navigate the ocean's depths.

Compared to us, sharks have extra-sensory perception (ESP), or senses in addition to sight, sound, or smell. On their head is a network of electroreceptors comprised of small gel-filled openings filled called the ampullae of Lorenzeni that are capable of detecting the weak electric fields of animals. Every animal produces a weak electric field due to the action of their nervous system and muscles. Hammerheads have taken their ability to detect electrical fields to an extreme because their head has evolved into a broad hammer shape to provide a wider area for the ampullae of Lorenzeni to detect small crustaceans buried in the bottom of the ocean.

Obviously, sharks are successful predators. They possess several additional adaptations to help them hone in and catch their prey. The Great White Shark ranks among the most well-known and feared predators in the world and for good reasons. Similar to other sharks, they possess the ampullae of Lorenzeni allowing them to detect the beating heart of their prey. They can smell tiny amounts of blood in the water. Great whites are dark on the bottom and light on the top, a color pattern known as counter shading, making them difficult to detect either above or below them. And of course, they have large powerful jaws full of sharp teeth capable of tearing flesh. It's easy to see why the Great White is an apex predator, at the top of the food chain. They can reach lengths upwards of 20 feet, however that pales in comparison to their extinct relative *Megalodon*. It has been estimated that *Megalodon* reached lengths in excess of 50 feet and most likely preyed on whales. About 2.9 million years ago, they went extinct for unknown reasons.

Not all sharks are top predators, in fact, the largest living shark, the Whale Shark often exceeds 50 feet in length. They are giant filter feeders, gently swimming through the ocean collecting small shrimp and fish. Part of the reason why they are able to reach such large sizes is because they feed closer to the base of the food chain where there is much more prey available.

The skates and rays represent the other major lineage of cartilaginous fish with their flattened bodies and gills located on their underside. Rays are kite-shaped with long tails, and the stingrays possess one or two stingers at their base. Rays are also live-bearers, but skates lay eggs called a mermaid's purse, which is often seen washed up on sandy beaches. Reaching 27 feet across, the manta ray is the largest ray, despite their large size they are often seen jumping out of the water. Manta rays are harmless filter feeders, often observed by divers.

Perhaps one of the most interesting rays is the sawfish, which actually bears little resemblance to other rays with elongated body and generally, more "shark-like" appearance. Despite their more shark-like appearance, their gills are located on their underside, similar to all other skates and rays. Sawfish have the most unique feature, an elongated saw-like rostrum possessing numerous ampullae of Lorenzeni, the electro-sensitive pores allowing sawfish to detect prey hiding under the sea floor. When they find their prey, they slash at it with their saw-like rostrum.

Bony Fish: an Evolutionary Success Story

Bony fish first evolved over 420 million years ago, about the same time as the cartilaginous fish. The rise of bony fish can be considered an evolutionary leap for several reasons. While they get their name because they form true bone made from a specialized type of cell called an osteocyte. Bone is different from cartilage in several ways, importantly bone is a dynamic tissue capable of growth and repair. Their true success, however, may have been caused by the evolution of the swim bladder. However, there has been some recent evidence that the swim bladder may have started as a primitive lung that allowed gas exchange with the atmosphere. That's correct, lungs may have evolved in fish before terrestrial vertebrates. Some scientists have hypothesized that fish evolved lungs because they lived in warm areas where dissolved oxygen levels could be temporarily depleted, or they simply required more oxygen for their increasingly active life styles. They think that the lungs were modified into a swim bladder allowing bony fish to control their buoyancy.

The evolution of swim bladders is another example of evolutionary serendipity because it helped fish in two important ways. First, it saved the fish energy from having to constantly swim. Recall that sharks lack a swim bladder and sink when they stop swimming. Second, it freed the fins from being used to maintain lift in the water and allowed them to be modified for maneuverability. Together, these modifications helped to make bony fish quite successful.

There are two major linages of bony fish, the ray-finned fishes and the lobe-finned fishes, which include the tetrapods. You are most familiar with the ray-finned fish, including clown fish made famous by the Disney film *Finding Nemo*, along with popular game fish such as tuna, salmon, snapper, grouper, and dauphin fish (popularly called Mahi Mahi to make it more desirable at market and fetch a higher price).

Diversity of lobe-finned fish was at its peak in the Devonian and declined by 365 million years ago. Today, there are only 9 known species of lobe-finned fish still living. However, their ancestors did give rise to the tetrapods, of which there are about 30,000 species. A rate of diversity similar to the ray-finned fishes. When you think of the most common species of fish, especially the ones at the grocery store or at the local pet store, they are usually ray-finned fish. They can be found just about anywhere there is permanent water, with the exception of extremely hot, saline, or acidic environments.

If you're like me, I'm interested in the biggest and the fastest animals. The fastest fish in the ocean are the sailfish (*Istiophorus*), which can reach top speeds of 68 miles per hour in water! The Bluefin Tuna can reach over 10 feet in length and weigh over a thousand pounds, yet it can reach speeds close to 50 miles per hour. Until recently, it was generally assumed that all fish were ectothermic (ecto=out, therm=heat), which means their body temperature is dependent on the temperature of their environment. We commonly refer to ectotherms as cold-blooded, but that can be misleading as many ectotherms are most active at the same temperature as a mammal or bird. In true scientific fashion, a scientist caught a small tuna and was amazed at how fast it could move its muscles considering how cold the water was. Muscles are made or proteins and are able to contract faster in warmer temperatures. Upon making the observation of the fast speed of the tuna's muscles, he measured the temperature of the fish to find that its muscles were actually 10 degrees above the temperature of the water. This one observation changed decades of thinking about fish and also illustrated the importance of curiosity, making observations, asking questions, and experimentally testing our assumptions.

More fish with ESP? In Africa there are a fish with a long "snout" known as elephant fish in the family Mormyridae. They are capable of generating weak electric fields that are used to "see" their surroundings and detect potential prey. The snout and body possess numerous sense organs similar to the ampullae of Lorenzeni in sharks. Some evidence exists that some elephant fish may actually communicate through their electric fields. What's amazing is that some elephant fish have a brain to body ratio that rivals humans, the large brain is required to handle all the extra sensory inputs. Electric eels and electric catfish are also capable of producing electric fields capable of completely stunning their prey, delivering a jolt of several hundred volts!

When we think of venomous animals, rattlesnakes almost always come to mind, but there are about 1200 species of venomous fish! There's a difference between a venomous and a poisonous animal; a venomous animal injects a toxin into an animal either through fangs (modified teeth), fins, or stingers, whereas a poisonous animal is toxic if you ingest it or absorb the toxin through the skin. Perhaps the most commonly known venomous fish are the lionfish (*Pterois*), a popular aquarium species. Their long spines are capable of injecting a neurotoxin that can cause severe pain and in rare cases lead to death. As a result, they have very few natural predators, a major reason why their numbers have greatly increased in the Atlantic where they have become an invasive species. Fins are not the only venomous part of fish. One small marine fish commonly found inhabiting tropical reefs, called the fang-tooth ed blenny (*Omobranchus ferox*) has fangs on their lower jaw which can be used in self-defense.

Ray-finned fish come in all shapes and sizes, dependent on their habitats. You can tell a lot about how a fish just by looking at its shape. Torpedo shaped tuna can maintain high speeds for hours as the bolt through the open ocean. Wheras the leafy sea dragon (*Phycodurus eques*) is the exact opposite, its fins have been highly modified to resemble sea weed to avoid detection by both predators and prey. Flounder have become flattened to hide under sandy bottoms to ambush their prey. Some of the scariest looking fish are the voracious looking denizens of the deep sea with their small bodies and enormous mouths full of long sharp teeth. There is not a lot of food in the deep sea, a meal might come along only rarely. That is why many deep sea fish have evolved huge mouths to catch large prey so they can go long times between eating. Some of these fish have formed symbiotic relationships with bioluminescent bacteria to attract prey.

With a long evolutionary history, ray-finned fish have also evolved some rather unusual mating habits. The males of some deep sea anglerfish are barely more than a degenerate sperm donor. Because it can be hard to find potential mates in the deep sea, the male will permanently latch on the female's skin for the rest of his life, fertilizing the female's eggs when needed. The bonding is so complete that even their tissues fuse together to the point where the blood vessels between them fuse providing the male with all the nutrients and energy he needs. We already know that clown fish, groupers, and wrasses can change sex; but in the sea horses, the male has a pouch located on his abdomen where he holds all the babies. Some fish will even brood their babies in their mouths, protecting them from predation.

External reproduction is the ancestral trait for vertebrates, but internal reproduction has evolved more than once in the vertebrates. The common guppy and mosquitofish throughout the US reproduce internally and give birth to live young. In fact, the least killifish (*Heterandria formosa*) of the southeastern US is the smallest live-bearing vertebrate in the world. In Chapter 6 Reproduction, I also discussed serial hermaphrodism, where fish change sex based on who's in their environment.

Lobe-finned Fish Gave Rise Tetrapods

It may be hard to believe, but the distant relative to all tetrapods, animals with four limbs that include amphibians, reptiles, birds, and mammals, evolved from a lobe-finned fish living about 375 million years ago. We aren't exactly sure what drove lobe-finned fish to colonize the land. Although, several good hypotheses have been proposed to explain why they transitioned from water to land.

The first hypothesis invokes competition, a force of nature that constantly drives the evolution of new species and innovations. Recall that the Devonian was the Age of Fishes, a time when fish were morphologically quite diverse and likely fueling intense competition. Any species capable of moving onto land would experience less competition with other fish. The second hypothesis proposes that the move to land was to exploit new resources. When fish were making the transition to land, there were no terrestrial vertebrates on land, but forests were present and arthropods had been there for millions of years. By moving to the land, there were many new resources available to them. In the end, the transition to land was probably a combination less competition from other fish and the move provided many opportunities to exploit new resources.

During the Devonian from 419 to 359 million years ago, the lobe-finned fish were in their heyday of diversity. Of all the lineages of fishes, they were the best suited to make the transition to land. So, how does a fish evolve into a tetrapod? A common theme in evolution is that natural selection acts to modify features already present amd repurpose them for new uses. The repurposing and modification of lungs, paired limbs, and a bony skeleton played a major role in the evolution of lobe-finned into terrestrial tetrapods.

About 380 million years ago, the transition to land likely began when a unique lobe-finned fish called a lungfish started spending part of their time at the water's edge and possibly some time on land. Their primitive lungs would help them obtain oxygen from the atmosphere.

Recall that lungs may have first evolved in fish. Ancient species of lobe-finned fish lived in slow moving or stagnant water. During times of warm temperatures, the water would lack sufficient oxygen for fish to be active or even survive. To make up for their oxygen debt, these fish may have evolved lungs so they could come up and gulp air to survive. The end result is that by having a lung already present, it made the transition to land easier because they already had means to breathe air.

Both ray-finned and lobe-finned fish have paired appendages, two pectoral fins and two pelvic fins. However, the fins of the lobe-finned fish are connected to the body by a single bone attached to muscles. The single bone is then attached to other bones further out in the fin. This arrangement was important because it helped to support their body on land. As they continued to evolve to life on land, their fins eventually evolved the universal configuration of one bone, two bone, lots of smaller bones, including a wrist allowing them to more easily walk on land. Even today, the bones in our own limbs still follow the convention of all other tetrapod of one bone, two bones, lots of bones.

Lobe-finned fish are also a lineage of bony fish, meaning they have a mineralized skeleton. In watery environments, the skeleton would provide sites for muscle attachment allowing fish to swim with much more power and speed. The bony skeleton was another important adaptation in fish that was helpful in the transition to land. Once on land, the first tetrapods had to deal with gravity, a force that is not very important in water. Fortunately, the bony skeleton was easily modified to become larger and thicker and therefore capable of supporting their bodies on land. It's almost as if the lobe-finned fish with their lungs, lobed fins, and bony skeleton were preadapted to living on land.

Today, only 9 species of lobe-finned fish are still living, including the coelacanth found in the Indian oceans and a few species of lungfish found in freshwater habitats in Africa and Australia. Despite lack of modern diversity in lobe-finned fish, their legacy as tetrapods continues today, rivaling their ray-finned cousins.

Amphibians Were the First Tetrapods

Most tetrapods are vertebrates capable of breathing air through modified lungs (although, there are a few modern lungless salamanders). The first were amphibians resembling a large salamander while retaining some fish-like qualities. Reminiscent of their fish origins, all of these animals had to return to the water where they reproduced externally. Although, the first fully formed amphibians appear in the fossil record about 370 million years ago, modern amphibians, which include three lineages, the frogs and toads, salamanders, and the worm-like caecilians first appear in the fossil record about 240 million years ago. This is about the same time as when the dinosaurs were first evolving and beginning to dominate the land. Today, there are approximately 7,000 species of amphibians, 90% are frogs and toads. Almost all amphibians must return to water to breed and still reproduce through external reproduction. However, as you can imagine, they are quite diverse in the details of where and how they reproduce in water.

Most amphibians return to water to lay their eggs. However, laying eggs in permanent streams poses a problem, they may face higher rates of predation from fish. Therefore, some frogs have come up with unique adaptations to prevent their tadpoles from being eaten. Glass frogs will lay their eggs on leaves over a stream where they will guard them. When it rains, the eggs will wash into the stream where they will develop into small frogs. A poison dart frog in South America called *Colostethus* will actually carry their tadpoles on their backs. Other poison dart frogs will lay their eggs in bromeliads, a common air plant found growing on trees high up in the canopy. Because of the abundant rain, these plants will often have water pooled in their folded leaves, providing a safe place for their eggs.

Some frogs and toads have been able to successfully survive even in the desert southwest of North America where water is scarce. During the summer monsoon months, you can slowly drive country highways at night (commonly known as road cruising), and if you're lucky, you will find spadefoot toads sitting in the middle of the road. Spadefoot toads get their name from a small spade on their hind legs that help them to burrow into the ground where they spend most of their lives. You may notice that there is little or even no permanent water anywhere in sight, leading you wonder how they are surviving in such a dry environment. These small toads remain burrowed underground until the onset of the summer monsoonal rains that occur from July through September. In low lying areas, temporary ponds may form in low lying areas. Most of these ponds will hold water a few weeks to a few months, depending on the amount of rain and their location. At the onset of monsoonal rains, spadefoot toads will come out, mate, and lay their eggs. The tadpoles are adapted to quickly morph into small adults before the ponds dry up. At the end of the monsoon season, they will burrow back into the ground and wait for the next monsoon season.

Perhaps the most bizarre adaption to protect their young is performed by the Darwin's frog (*Rhinoderma darwinii*) found in South America. After mating with a female, the male will swallow the eggs and hold them in his vocal sac until. Once the tadpoles grow into small frogs, the father will cough them up one at a time. Even more amazingly, he will produce a nutrient rich liquid, similar to mammalian milk to help nourish his progeny. In Australia, there were two species of gastric brooding frogs in the genus *Rheobatrachus*. The females would incubate the larvae in her stomach until they were older, and similar to the male Darwin's frog, she would cough them up when they were small frogs. Unfortunately, both species of gastric brooding frogs are most likely extinct due to habitat loss and the Chytrid fungus. However, there is an attempt by Australian scientists called the "Lazarus Project" to bring back the species by cloning them.

Frogs and toads have a unique body plan, their hind legs and pelvis region are specialized for jumping or hopping. In North America, the green tree frog (*Hyla cinerea*) can jump upwards of 150 times its body length, second only to the flea. Not every frog can jump so far, and toads merely hop, which would normally make them an easy catch for a predator. To protect themselves, some frogs and toads are poisonous, producing toxins in their skin making them taste bad. Toads have a large warty-looking bump behind their eyes, called a parotid gland, which produces a neurotoxin called bufotoxin. It is potent enough to kill most animals interested in eating it. In Central and South America you can find the small, brightly colored poison dart frogs in the family Dendrobatidae. They produce toxic substances called alkaloids, which can be quite toxic when ingested.

Salamanders are a unique group of amphibians dating back to the middle Jurassic, about 170 million years, which ironically resembles the first tetrapods. The largest amphibian is the Chinese giant salamander (*Andrias davidianus*) found in China, which can reach a length upwards of six feet. They have large teeth that often inflect harsh wounds when the males fight over their territories. One family of salamanders called the sirens (family: Sirenidae) appear eel-like and actually lack a pelvis and hind-limbs. As adults, they possess external gills, lack eyelids, and are often found burrowing in the mud. In North America are the lungless salamanders in the family Plethodontidae, a lineage that has lost their lungs. Instead, they respire across their skin and through the tissue linings of their mouth. In the eastern United States, they can become quite numerous with population estimates of 1.8 billion in a single forest. Their small size and slow metabolism allow them to reach such high numbers. Salamanders have one very interesting ability that is of major interest to us, they can regenerate their limbs if they lose them.

Amniotes: Adaptations for Living on Land

Sometime around 312 million years ago, a group of tetrapods evolved several useful adaptations greatly improving their ability to survive on the land by reducing their dependence on returning to the water. Amphibians have retained the fish-like characteristics of external reproduction in the water. But, this new group of animals evolved the ability to reproduce internally and place the embryo in a protective, water-filled structure called the amniotic sac. The evolution of these two features, internal reproduction and the amniotic sac protected by a hard shell, allowed tetrapods to lay their eggs on land. Shortly after the appearance of amniotes about 300 million years ago, they split into two lineages.

One lineage would evolve into reptiles, a very diverse assemblage of animals including lizards and snakes, turtles, pterosaurs, ichthyosaurs, mosasaurs, plesiosaurs, crocodiles, dinosaurs, and birds. The other lineage of amniotes would eventually evolve into modern day mammals. Amniotes are a lineage whose evolution was largely driven by the spread of dry environments. But, they have also been able to repeatedly return to both marine and freshwater habitats.

Reptiles
Vertebrates Conquering the Land

The ancestors of reptiles and mammals began to diverge a little over 300 million years ago. Despite the vast amount of time since their divergence, mammals and reptiles still reproduce internally and produce a water-filled amniotic sac to protect their embryos. Today, reptiles are a diverse lineage even though their heyday may have ended 65.5 million years ago with the extinction of most dinosaurs and their relatives. Today, we typically classify modern reptiles in two broad groups, the Archosaurs and the Lepidosaurs, although we aren't quite sure where turtles fit in. Ironically, Archosaurs means "ruling lizards", but they aren't true lizards. Living Archosaurs include birds and crocodilians, but there are many extinct forms including the dinosaurs and the flying pterosaurs. Lepidosaurs mean "scaled lizards" and include both snakes, lizards, and the primitive looking tuatara. There are other extinct reptiles from the Mesozoic including marine forms such as ichthyosaurs which resembled modern dolphins, plesiosaurs, and the very large mosasaurs like the one seen eating a large shark in the movie *Jurassic World*.

Most reptiles are ectothermic, meaning their body temperature is the same as the environment, or they do not produce metabolic heat. Being ectothermic, reptiles are somewhat limited in their distribution by colder climates. Reptiles also have scales, an evolutionary adaption that helps prevent water loss across the skin. There are unique features of their skulls that are important in their classification as well, including the shape of the jaw, ear bones, and the number of openings in the skull. Most reptiles also have a three chambered heart, although birds and crocodiles have four chambered hearts that prevent the mixing of oxygenated and deoxygenated blood. The structure of the heart is also a key feature for quickly moving material throughout the animal and allowing them to evolve into large animals.

Turtles

Perhaps one of the most recognizable reptiles are the turtles with their hard shells. In fact, turtles are so distinct that scientists have had difficulty understanding their relationship to other reptiles. Just how and when did turtles evolve their shells? Based on fossil records, their origins date back to the Jurassic Period 240 million years ago when they last shared an ancestor with other reptiles. That makes them an older group than dinosaurs or crocodiles. Today, there are about 327 species, and many are quickly becoming endangered due to habitat loss and over harvesting. Turtles live a long time, with giant tortoises living more than 200 years!

Many turtles have a close association with water. Freshwater turtles can be easily seen basking in the sun along the banks of streams and rivers. Sea turtles have all but completely returned to the sea. The largest species of turtle is the Leatherback, which can live upwards of 80 years. Its evolutionary roots date back over 100 million years ago to a time when the dinosaurs were dominant on land. These turtles can dive down to 4800 feet and hold their breath for 85 minutes. Leatherback sea turtles are found throughout the world's oceans eating jellyfish. Unlike other reptiles, leatherbacks exhibit a small degree of endothermy. Because of their large size, they are able to maintain body temperatures above their environment, enabling them to survive in colder waters.

The remote ancestor to leatherbacks and all other sea turtles once lived on land and returned to the sea only to return to lay their eggs. Today, sea turtles often make long migrations swimming over 3700 miles one way to lay eggs on the same beach where they hatched. It's been hypothesized that the reason why they make such long migrations is due to continental drift. A hundred million years ago, their ancestors didn't have to swim as far, but with the continents moving apart at the rate of an inch per year, that distance has become quite large over time. So each year, they swim a little further to lay their eggs.

Lepidosaurs
The Scaled Reptiles

Lizards, snakes, and the Tuatara belong to the Lepidosaurs (lepido = scaled saur = lizard). Despite not having legs, snakes are closely related to lizards. In fact, the loss of limbs in lizards has occurred multiple times. Worldwide, there are about 3,400 species of snakes living in a variety of habitats. Sometimes, they can be quite common, but are rarely seen making it difficult to estimate their populations.

Snakes are unique predators that have evolved several senses to find, catch, and swallow their prey. Snakes are often seen flicking their tongues to detect odors in the air and to taste the ground. Pit vipers in the Americas possess small pits below the eye that are capable of detecting heat. When it comes to catching prey, some snakes are constrictors. They can quickly grab their prey and then wrap around them multiple times. Each time the hapless prey breathes, the snake will tighten down, eventually suffocating their prey. Snakes have modified jaws allowing them to swallow prey that are bigger around than they are. Contrary to popular belief, the jaws are not unhinged, they just aren't rigidly attached to the skull. The lower jaw bone is connected by ligaments allowing them to move independently, unlike our lower jaw where both sides move together.

Snakes are often feared by people, most likely because some are highly venomous. There are two main families of venomous snakes, the Elapids and the Vipers. The Elapids include cobras and coral snakes, which are found on most continents. In the US, the coral snake has a very potent neurotoxic venom that can cause respiratory failure within a few hours. Vipers are found in the western and eastern hemisphere. In the U.S., they are represented by the pit vipers which include several speices of rattlesnakes (*Crotalus*), copperheads (*Agkistrodon contortix*), and water moccasins (*Agkistrodon piscivorus*). Their venom can be neurotoxic or cytotoxic depending on the viper species and the type of prey. The rattle is formed by a series of keratinized interlocking scales that rattle when shaken. Rattlesnakes retain their eggs inside until they hatch, giving birth to live young.

Lizards are closely related to snakes and can be quite common in some habitats. As of 2015, there are about 6,145 species found on all continents except Antarctica. In the Western Hemisphere, lizards in the genus *Anolis*, have over 400 known species, making it the largest genera of amniotes in the world. Anoles are known for their dewlaps, a colorful patch of skin that they can extend and retract, which the males use to attract females or establish territories. Different species often have dewlaps of different color and sizes. Some anoles, including the Carolina anole in the south eastern U.S. are capable of changing their skin color to match their surroundings. Recent expeditions to the tropics, led by Dr. Steve Poe from the University of New Mexico, continue to discover new species of Anoles.

Another very successful lineage of lizards are the geckos, of the 6145 species, 1500 are geckos. Most geckos are nocturnal; their eyes are 350 times more sensitive to light than ours. They have highly specialized toepads that allow them cling to variety of surfaces. In some cases, they have been observed walking across ceilings. Some species of geckos have done well in human environments where they can often be observed at night on walls where they prey on insects attracted to the lights.

The largest lizard in the world is the Komodo Dragon, a type of monitor lizard capable of growing up to 9 feet long and weighing about 150 pounds. It's unclear why this lizard is so large, but it may be relict of an older group of large lizards that went extinct about 15,000 years ago. They are ambush predators taking invertebrates, birds, and mammals, including small deer. Interestingly, Komodo Dragons have been known to hunt together, which is unique among reptiles.

Archosaurs
The Ruling Lizards

It may seem odd to include dinosaurs, birds, and crocodiles into one lineage. After all, birds and crocodiles don't seem to have much in common. Birds are instantly recognizable by their feathers, wings, and bills lacking teeth. Alligators, on the other hand, are large and sprawling, with powerful jaws full of teeth. But, if you look closer, they both have scales and four-chambered hearts. Additionally, birds and alligators share several behaviors; they make nest to lay their eggs and "sing" to their mates. Because these behaviors are shared in both groups, it is generally accepted that dinosaurs also made nests and sang to their mates.

There are about 24 species of crocodilians in the world, found mostly in the tropics. They have a nostril on top of their head so they can breathe while remaining mostly submerged in water. Most of them are ambush predators, quickly launching themselves at their prey. Once they catch their prey, crocodiles will drown it by repeatedly rolling in the water. The Saltwater Crocodile (*Crocodylus porosus*) is the largest living reptile, males have been known to reach 20 feet and weigh over 2,000 pounds. Capable of clamping down on their prey at 3,690 pounds per square foot, they also have the largest bite force of any terrestrial vertebrate and rival adult Great White Sharks.

For nearly 165 million years of the Mesozoic Era, the dinosaurs were the dominant land animal. So far, paleontologists have identified about 700 species of dinosaurs indicating that they were very diverse in their size and habits. Based on unearthed fossils, dinosaurs ranged in size from the tiny *Microraptor* that was barely over a foot long to the massive *Titanosaurs*. These large herbivrous dinosaurs are estimated to have been 130 feet long and weighing up to 90 tons. The first dinosaurs first evolved in the Triassic about 230 million years ago and walked on two legs. Some were carnivores and others were herbivores.

Whether or not dinosaurs were active endothermic animals is still up for debate. Although, there is some evidence indicating that they were fairly active animals. By comparing the traits of their closest living relatives, the birds and crocodiles, we assume that dinosaurs had a 4-chambered heart, which is an important adaptation for large active animals. In recent years, new fossils have shown that dinosaurs evolved feathers long before the first birds evolved. We aren't quite sure why dinosaurs evolved feathers, but it probably wasn't for flight, that would come later. It is though that feathers could have been used for insulation in colder temperatures, or they could have been used in displays to attract mates. Feathers are used by birds for both those functions today. Also, modern birds are endothermic, a necessary condition for powered flight. Endothermic birds, the presence of feathers, and a four chambered heart is good evidence supporting that at least some dinosaurs may have been endothermic with active lifestyles. Although, the case is not closed on our understanding of dinosaur physiology, paleontologists are still actively searching for additional evidence in the fossil record.

In a recurring theme of evolution, a feature evolves for one purpose and then is repurposed for other uses. Evolving first in dinosaurs, feathers were likely used for insulation or to attract mates, but not flight. Starting about 145 million years ago, feathers were modified for flight as the first birds evolved from the Theropod dinosaurs. Today, birds use feathers for all three purposes, flight, insulation, and to attract mates. Because birds are direct descendants of dinosaurs, it can be said that the dinosaurs never really went extinct. We see their descendants flying around us every day!

With 10,000 species of birds today, their diversity is simply amazing. The smallest species of bird is the tiny Bumblebee Hummingbird, which is not much bigger than the eraser at the end of your pencil and weighing less than an ounce. In fact, they are so small that their size is about the lower limit of for endotherms. Their metabolism is so high that they could die within a few hours of not eating. To get around this problem at night when they are unable to forage, hummingbirds go into torpor when they lower their body temperature to conserve energy. At the opposite end of the size spectrum is the Ostrich, a very large flightless birds of an ancient lineage. They can stand almost 9 feet in height and weigh up to 200 pounds, and have been clocked running at 50 mph.

Although most birds can fly, there is a wide range in their flying abilities. The Greater Roadrunner of the American Southwest gets its name because it almost always seen "running", in fact it's rare to see one fly. When they do fly, it is mostly to avoid predators and they don't fly very far. The fastest birds in the air are the Peregrine Falcons, they've been clocked at speeds of 240 mph while dive-bombing their prey. Humming birds have the fastest wingbeat, about 80 beats per second and they are the only bird capable of flying backwards. A denizen of the open oceans is the Wandering Albatross with very long and narrow wings built for soaring. They can ride the air currents for days at time as they wander the oceans. Although they are graceful in the air, they have to get a running start to fly.

Birds are known for their songs. While most birds can make some sound, even if it is little more than grunt, the song birds take it to the next level. Songbirds are are a sinle lineage called passerines and account for about half of all bird diversity. The vocal cord in birds is called a syrinx and is analogous to our larynx. In some song birds, the syrinx is split in two, allowing birds to make two sounds at once. This is especially noticeable in the Wood Thrush and Hermit Thrush, two brownish birds commonly heard singing their melancholy sounds in forest throughout North America in the summer.

The bills and feet of birds are also adapted to very different lifestyles. Web feet has evolved independently in penguins, gulls, pelicans, and ducks, all capable of swimming on and in water. Likewise, the bills of these birds are also quite different. Pelicans have a pouch-like beak that they use to scoop in lots of water and food at the same time. Penguins have much smaller bills for catching fish directly. And ducks have wide bills for filtering plant material from the water. Although, some ducks have a long and narrow serrated bill to help them catch fish. Songbirds are also quite varied in their diets and thus the shape of their bills. Finches and cardinals have thick bills for cracking seeds, whereas warblers have much smaller and thinner bills for catching insects. Often seen on sandy beaches and mudflats, sandpipers have tubular bills used to probe the soil for small invertebrates. They have bills of varying length so they can reduce competition between the species by foraging at different depths and habitats.

All birds lay eggs, not a single species gives live birth. Flight is an incredibly metabolically demanding activity, and birds have evolved numerous adaptations to accommodate their lifestyle. They are endothermic, generating their own body heat, which is necessary for fast muscle contractions. Their bones are hollow to conserve weight, and they are fused for added strength. One really interesting adaptation is their lungs, they are unidirectional, meaning air only goes one way through their lungs to maximize oxygen uptake. Another unique feature of birds is that they live a long time compared to mammals of similar size, and they rarely show signs of aging in the wild. In a remarkable series of photographs, a scientist caught the same seabird about once a decade over a period of 50 years. In each photograph, the scientist can be seen aging, but the bird always looked the same. To understand the aging process, scientist have begun to study birds to determine how they live so long.

Mammals
Bigger, Faster, and Smarter

Mammals share a common ancestor with the reptiles, we are both amniotes. Surprisingly, the ancestors to mammals were the dominant terrestrial vertebrates until about 251 million years ago when a massive extinction wiped most of them out. Twenty million years later, the dinosaurs evolved to dominate the land, successfully outcompeting our mammalian ancestors. But, the ancestor to mammals continued to evolve in the shadows of the dinosaurs. Over time, they evolved mammary glands and extensive parental care. Mammals evolved endothermy so they could remain active at night so as not to compete directly with the dominate reptiles. To stay warm, they evolved hair to insulate them from heat loss, and a four-chambered heart to increase their level of activity. Early mammals also developed a keen sense of smell to help them find food at night. By 205 million years ago, mammals had modern appearance and would be easily recognizable today. For the next 140 million years, most mammals remained small and nocturnal. Then, 65.5 million years ago, the dinosaurs went extinct, creating an opening for the mammals to diversify.

Mammals are a diverse group with a long evolutionary history. Mammals had been around for 140 million years when the dinosaurs went extinct, but by the beginning the Cenozoic Era 65.5 million years ago, they rapidly diversified to fill the ecological spaces previously held by dinosaurs. Although, for the first few million years after the dinosaur extinction, birds were also diversifying and competing with mammals for dominance. However, mammals eventually secured their dominance in terrestrial ecosystems, including the top predators.

As mammals diversified, some returned to the oceans, others took to the air, and some stayed in the trees only to come down in the last 5 million years and walk upright. The features mammals evolved during the Mesozoic Era, including endothermy, hair, and a four-chambered heart, were important factors priming mammal for their eventual success. Today, there are approximately 5400 species of mammals that reside on every continent except Antarctica, and they are found in the oceans, on the land, and in the air.

There are three major lineages of living mammals, monotremes, marsupials, and eutherians. Monotremes are an ancient egg-laying lineage of mammals that include the Platypus and Echidnas of Australia. They are still considered mammals because they have hair and produce milk for their young. About 160 million years in the Jurassic, Eutherians split from the marsupials, forming a new mammal lineage where the fetus is carried in the uterus and connected to the mom by a placenta.

The Marsupials include about 334 extant mammals with the young being born at a very early age of development. Many species carry their young in a pouch. Basically, marsupials come in two major lineages, the

American marsupials and the Australian Marsupials. Although, fossil and molecular evidence indicate that Marsupials may have reached Australia by way of Antarctica from South America over 50 million years ago during the break-up of Pangaea. Today, about 70% of all living marsupials are found in Australia, with another 100 species living mostly in South America. Marsupials include several well-known species including koalas, kangaroos, possums, opossums, bandicoots, and the Tasmanian devil.

Molecular and fossil evidence suggests that a single species may have diversified into the all the species living today on the isolated Australian continent. One of the best known Australian marsupials are the kangaroos, which are the largest animals to move by hopping on large hind legs. Another well-known group of marsupial are the tree-dwelling Koalas who spend their time eating the leaves of eucalyptus trees. Interestingly, they have one of the smallest brains relative to their body size for any mammal. It takes up only about 60% of the brain case. The rest of the space is taken up by fluid, possible to protect their brains if they fall. The Tasmanian devil is the largest carnivorous marsupial and is about the size of a small dog.

Eutherians, a name which means "new beast" are by far the most diverse lineage of mammals with over 5000 named species. They, like the reptiles, have been able to take flight and return to the oceans. Eutherian mammals range in size from tiny shrews weighing only a few grams to the Great Blue Whale weighing over 170 tons. They include many of the commonly encountered and well known mammals including deer, elk, cows, bears, cats, dogs, rodents, bats, sloths, elephants, primates, and cetaceans along with strange looking scale-covered pangolins and armadillos. The two most diverse groups of mammals are the bats and rodents.

On an African safari, you get to see many mammals on the Serengeti plains. Many of these animals are ungulates a large group of even-toed mammals including cattle, pigs, giraffes, rhinoceroses, hippopotamuses, and wildebeest. In the U.S., we have our own ungulates, including mountain goats, bighorn sheep, elk, and deer. In the Serengeti, several species of large cats hunt many ungulates by ambushing them. To defend themselves, the ungulates, such as the wildebeest, form large herds to minimize their risk of being caught. Other mammals, such as water buffalo, have evolved to become very large making it difficult to be preyed upon, while others remain small and agile to outrun the predators.

Ungulates belong to a group of animals that often produce antlers or horns. Antlers are bony extensions of the skull grown by deer and elk, and are shed each year. Cows, goats, and sheep have horns, which are keratinized structures that are not shed. Instead, they continually grow throughout the animal's life. Interestingly, even-toed mammals related to elk, deer, and hippos share a common ancestor with whales.

Whales are cetaceans and include the largest animals, at least by weight, to have ever inhabited the Earth. The Blue Whale reaches lengths of 100 feet and may weigh upwards of 191 tons or 382,000 lbs. To achieve such large size, they feed on small shrimp called krill which are at the base of the food chain. Humpback whales are known for their intricate songs, which are sung by the males. They have been known to begin a song, migrate thousands of miles, return to the same spot and pick up singing where they left off months earlier. Whales may share little resemblance to other mammals, but the evidence of their relationship lies in the fact that their front fins maintain the same bone pattern of one bone, two bones, and lots of bones as other tetrapods. They also have live young and feed them milk. In some cases, up to 100 gallons per day for larger whales.

Not all cetaceans are giant filter feeders. The dolphins, orcas, and sperm whales possess teeth and are active predators. Dolphins form pods, sometimes with hundreds of individuals. The Bottlenose Dolphin is commonly encountered along the Gulf and Atlantic coasts in pods ranging from 10-30 members. They often hunt for schools of fish as a group coordinating their efforts using echolocation. Dolphins have very large brains, with a brain to body size ratio similar to humans.

Mammals have also evolved to become excellent predators on the land, although none have achieved the same size as the largest dinosaur predators. Many of the most well-known mammalian predators are in the order Carnivora, which includes, wolves, cats, raccoon, bears, hyenas, otters, and badgers. These predators have evolved different methods of catching their prey. The cats in the family Felidae are mostly ambush predators of different sizes that correspond to their prey. Lions are the largest cats weighing up to 500 pounds. Hunting in small groups called prides, they will take on larger mammals such as wildebeest, zebras, buffalo. Cheetahs are much smaller, relying on their fast burst of speeds reaching 60mph, they outrun their prey. Wolves and wild dogs will chase their prey over longer distances slowly wearing it out.

Bats are the second most diverse group of mammals in the world. They evolved flight independently of birds, as can easily be seen by differences in their wings. Rather than using feathers, their wings are made by skin stretched between their digits. Most bats are nocturnal and can be seen flying at dusk. Rather than rely on eyesight for navigation, they use a form of sonar by emitting high pitched sounds that bounce off their surroundings. As they approach their prey, they rapidly increase the number of "clicks" to hone in on their prey. Bat flight may appear erratic; but in fact it's quite precise as they scoop flying insects with their tail. Bats are voracious predators, consuming thousands of insects every night. Although, not every bat eats insects. In Central America, you can find the vampire bat, it has modified teeth for cutting the skin of large mammals so it can drink their blood. The largest bats, also called flying foxes, are fruit eaters with a wingspan of 5 feet.

Rodents are the largest order of mammals with about 2270 known species accounting for about 40 percent of all mammals. They include many familiar mammals, such as mice, rats, and Guinea pigs, squirrels, chipmunks, prairie dogs, porcupines, and the capybara. You can find rodents on every continent, except Antarctica. They range in size from small mice to the 115-pound capybara of South America. Many species of rodents, such as prairie dogs, form large social groups and are vitally important to their ecosystems.

Sometime near the end of the dinosaur's reign, a small arboreal mammal evolved forward facing eyes and fingers with fingernails rather than claws. Forming complex social groups, these early primates began to develop larger brains, and for millions of years, they remained in the trees. Then about 5 million years ago, the climate began to change and the forests receded as grasslands expanded across eastern Africa. Adapting to the changing landscape, our primate ancestors came out of the trees and onto the grasslands. They began to walk upright, possibly because it made it easier to spot predators, or because walking on two feet was a much more efficient way of moving. Either way, it freed their hands for other purposes, notably making tools. Eventually, modern humans evolved around 200,000 years ago in Africa; a date that has been supported by both fossil and DNA evidence. Additional fossil and DNA evidence indicates that populations from Eastern Africa migrated and populated the rest of the world. In the last 10,000 years, humans have been incredibly successful as we have developed new technologies that have aided in our survival and proliferation in practically every part of the world. Today, there are 7.3 billion humans on the planet, making us the most abundant large mammal on the Earth.

Chapter 11

Life Interacts
with the Environment

The day to day ecological interactions between individuals
and their environment is natural selection
driving adaptive evolution.

Introduction

No living organism is an island existing by itself, isolated from all other life. Every organism must interact with the living (biotic) and the non-living (abiotic) parts of their environments. Ecology is the study of the relationships between living organisms and their environment. The scope of ecological studies is quite vast. At its smallest scale, the focus may be limited to studying populations of a single species. In contrast, large scale ecological studies focus on global patterns of diversity and productivity.

Ecology, like any branch of science starts by asking questions. Why do we find different plants and animals as we move across the landscape? Why are the western forests of North America dominated by pines, spruce trees, and other conifers, while the eastern forests are dominated by deciduous trees? Why are Greater Roadrunners common in the southwest, but not in the eastern U.S.? To answer these types of questions, ecologists rely on natural history, an understanding of the distribution of plants and animals to study their relationships to each other and their environment. Natural history is mostly an observational science relying less on experiments and hypothesis testing.

Every species has a geographical range where you can find it, in some places they may be more common than others. Ecologists define a species place in an ecosystem as its niche. A species niche is determined by its tolerances to physical conditions in the environment and biological factors, both of which work together to determine where you may find a certain species. Additionally, the deep history of the Earth and continental drift has also played a significant role in where we find certain types of animals.

The connection between ecology and evolution is important, the species we see today are the result of selection in the past. It's also important to understand that every adaptation comes at a cost, a trade-off in that you do something else less well. As a result, there are no perfectly adapted species. That's why you wouldn't find a desert adapted lizard thriving in a rainforest.

The Importance of Natural History
For Understanding Ecology

Ecology begins with an understanding of natural history, a field of study that relies on observations to determine how organisms in a particular area are influenced by their physical surroundings and other organisms. Birds are among the best-studied organisms in the world as legions of birders have contributed millions of observations of bird sightings to a single data-base called *eBird*, which is run by the Cornell Lab of Ornithology. Therefore, we know a lot about the natural history of birds.

For example, if you're a birder, then you may know to look for Black Oystercatchers on rocky outcroppings along the Pacific coast, or you may search for a Grace's Warbler in the canopy of a Ponderosa pine tree in the Western United States in the late spring through summer. Likewise, you may know that you will only find Limpkins, a medium sized brown bird, in Florida where apple snails are abundant and it doesn't get too cold in the winter. There is also seasonal variation in the birds you will find in a certain region. In the winter, ducks and geese will migrate southwards to overwinter in the southern states to avoid freezing water. When the temperatures rise in the spring, they migrate northward returning to their summer breeding grounds. Some birds Including, Greater Roadrunners, White-winged Doves, Western Scrub Jays, and House Finches are year-round residents in many western states. Sometimes, I go to the beach on the spring high tide to watch thousands of shorebirds congregate at the water's edge to eat the eggs of horseshoe crabs.

From numerous observations, we can build geographical and seasonal range maps of individual species. We can also achieve increasingly more accurate range maps for a particular species by further refining them based on suitable habitat. If you've ever picked up a where-to-go birding guide, it helps you locate hard to find species by sending you to the right habitat at the right time of year. To further improve your chances of finding rare birds, a good field guide may include specific behaviors and habits of the bird in question. For example, a Brown Creeper is almost always seen on the trunks of trees working their way up towards the crown. A Spotted Towhee is almost always on the ground in thick underbrush scratching the leaves searching for a meal. A Vermilion Flycatcher can be seen from a distance sallying for insects, quickly returning back to its perch.

These types of natural history observations are not limited to birds, naturalist have been collecting observations on many types of plants and animals. For example, the ground-dwelling whiptail lizards (*Aspidoscelis*) are active on the ground when it is hot. Different species of anoles may be found on grass, on small limbs, on the trunks of larger trees, or high up in the canopy. Plants too have different distributions. The columnar cacti of the Sonora desert aren't found in the drier Chihuahua desert, and they aren't found to the north or in higher elevations where it gets too cold. If you've ever driven into the high country of the Rocky Mountains from the eastern plains, the terrain transitions from grasslands, to juniper and pinyon, to Ponderosa pines that eventually give way to mix coniferous forests dominated by fir and spruce trees. Above 11,000 feet, the trees abruptly end at the tree line where the alpine tundra begins.

The distribution of plants and animals described above is determined by both the physical environment and their interactions with other living organisms. Recall that no organism is an island. Therefore, living organisms interact with the non-living (abiotic) and living (biotic) part of their world to survive. Over the course of evolutionary history, the number of species has continually increased, thus increasing the types of ecological interactions and the complexity of ecosystems. Where natural history is mostly observational based, ecology is often more question driven relaying on experimentation to better understand how ecosystems are structured.

An important concept in ecology is the niche. The niche of a species is where and what that species is doing in an ecosystem. But, what are the factors that determine a species' niche? We can break the niche of a species into the fundamental niche and realized niche. In broad terms, the fundamental niche is where a species can be found based mostly on the physical environment or abiotic factors, including climate, disturbance, soil type, salinity, *etc.* The realized niche is where you may actually find a species based on biotic interactions, including competition, predation, diseases, symbiosis, *etc.* The realized niche is always smaller than the fundamental niche for any species.

The Physical Environment
Determines the Fundamental Niche

The fundamental niche of an organism is determined by abiotic factors that limit where a species can live based on their physiological tolerances. The abiotic factors that affect the distribution of different species arise from the physical environment. Some of the important factors include the climate, types and frequency of disturbance, soil types, nutrient availability, light availability, and salinity (think marine versus freshwater ecosystems). Many of these abiotic factors also vary over time. For example, the climate of temperate regions may experience cold wet winters paired with hot summers. In the southwest where monsoonal rains are important, the first half of the summer is dry and the second half is wet from frequent thunderstorms.

A species' physiological tolerance to its environment places limits on where you find it in its geographical distribution and the time of year you find the organism. Freezing temperatures place a major limitation on the distribution for many animals. For example, ducks can easily withstand freezing temperatures because their feathers are great at insulating them. However, they must migrate southward in the winter because the ponds where they breed and forage in the summer freeze over in the winter. Tropical and subtropical organisms could grow well during the summer in most northern latitudes; however, freezing temperatures in the winter will kill them. That is why tropical species remain confined to the tropics. If you are a freshwater fish, then you can't swim into the ocean because the higher salinity would cause you to lose too much water. Likewise, a marine fish swimming up a river would absorb too much water.

Climate is often one of the most important abiotic factors determining where a species can live. However, determining the fundamental niche of a particular species can be difficult because several abiotic factors could be important and vary over time. Additionally, biotic interactions, such as competition, can further restrict a species range. Below, I focus on climate and disturbance determine the fundamental niche of a species.

Climate

Climate is at the heart of determining the geogrpahical range for many species because of physiological tolerances. Climate can defined as the pattern of weather conditions in an area averaged over a long period of time. Climate is the average weather for a region. It includes the average yearly precipitation and average temperatures; along with seasonal variations including winter lows and summer highs, or the presence of dry versus wet months. Weather is basically the high and low temperature plus the precipitation for any given day. Weather is what you get today or tomorrow.

Let's take the Ponderosa Pine, one of the most common trees in the western United States, as an example to understand how climate is influential in determining where a species can live based on its tolerances to the environment. If you live at the base of the Rocky Mountains, Ponderosa pines are largely absent and the ecosystems may be dominated by grasses or shrubs. However, if you were to drive or hike into the mountains, you would quickly notice these large pines once you reached about 7500 feet in elevation, give or take a few hundred feet. At this elevation, the average temperature is a little lower and the annual precipitation also increases as you go up a mountain. The connection here is that as you go up the mountain, the climate changes and becomes favorable for Ponderosa pines. At lower elevations, the climate is not favorable for Ponderosa pines, it becomes too hot and dry for this tree, so its distribution is limited by climate.

Sometimes the climate for a particular region can be complex and the organisms present must be adapted to changing conditions. For example, rainfall can vary dramatically throughout the year. Nowhere is this truer than in the desert southwest of Arizona and New Mexico. The first half of the summer is typically hot and dry, whereas the second half receives monsoonal rains starting in early July and ending in September.

Plants and animals have adaptation allowing them to coincide their life cycles with the onset monsoonal rains. For example, spadefoot toads and tadpole shrimp have quick development times. Spadefoot toads will grow rapidly as tadpoles and emerge not fully grown before the ephemeral ponds dry up. After the rains, they will use specialized spades on their hind legs to burrow deep into the ground to wait for the next monsoon season. Small primitive looking tadpole shrimp will hatch once their eggs become immersed in water, grow into adults, mate, and lay eggs all within about two weeks during the monsoon season.

In temperate regions, there are distinct seasons where summers are much warmer than winters. Contrary to popular belief, summer and winter climates are not based on the distance of the Earth from the sun, but on the angle of the hemisphere to the sun. The Earth is tilted on its axis changing the angle of incoming light as it orbits the sun. Winter occurs in the northern hemisphere when it is tilted away from the sun and summer occurs when it is tilted towards the sun. If the northern hemisphere is pointed away from the sun, then the incoming light is arriving at an angle so that less energy is hitting the Earth. When combined with shorter days, the temperatures drop, reaching their coldest temperatures typically in January. Because of the large temperature differences between summer and winter, organisms living in temperate regions have evolved numerous adaptations to deal with the cold or migrate to warmer regions.

Plants and animals respond in different ways to the seasonal changes in temperature. Deciduous trees will lose their leaves, completely shutting down photosynthesis for the winter. Once temperatures warm in the spring, they will once again grow new leaves. There are several benefits to losing leaves during the winter, one is that it avoids damage to the leaves if they freeze. In areas where snow accumulation is possible, the plants won't accumulate as much snow and ice on their branches that could potentially damage the tree by breaking limbs. Also, plants lose water through their leaves, if the ground were to freeze, the available water would become near zero, and the plant could face desiccation even with plenty of water in the ground. One last advantage is that plants would face less herbivory in the winter with no leaves. A disadvantage of being deciduous is that plants lose energy when their leaves are shed. Also, a deciduous could be a disadvantage when competing with an evergreen tree. Deciduous trees must grow new leaves to begin photosynthesis, whereas an evergreen can immediately begin to grow when conditions are favorable.

Some mammals will respond to colder temperatures by going into hibernation, allowing their body temperature to drop so they don't have to spend as much energy maintaining a high body temperature. Large mammals like Brown Bears will feed all summer long, building up large fat stores that they slowly use over the long winter. Hibernating mammals have brown fat that specializes in producing body heat by burning up their fat storage.

Birds are known for making long migrations, but many people are unaware of how many birds actually migrate in the spring and fall. Birds are among the most active animals on the planet with very high energy demands. To supply their energy, birds require energy-dense foods, that's why so many of them are predators, or seed eaters. There are very few birds that can survive solely on eating leaves. If they do eat plant material, they are eating mostly seeds and fruits. During the winter, insects become quite scarce, especially in areas where it routinely freezes. To follow resources, many birds will migrate south, tracking warm weather and abundant food sources. In North America alone, it's been estimated that approximately 5 billion birds migrate to Central and South America in the winter and then back to North America in the spring when it's warm. Shorebirds that nest in the Arctic Tundra migrate south to avoid the 24 hours of darkness and bitter cold of winter. One shorebird, the Black-tailed Godwit, will migrate from Alaska to New Zealand in one trip, traveling 8,000 miles and losing over a third of its body weight along the way!

The location of continents, oceans, mountain ranges, and prevailing wind patterns also play a role in the climate of a region. The atmosphere and the oceans move energy away from the tropics, towards the poles. Large atmospheric circulation patterns known as Hadley cells move air from the tropics to about 30° latitude north and south. The air near the equator is warm and humid, as it rises, it cools and drops the moisture as rain. That's why the tropics receive so much rain and tropical rain forests are confined to regions near the equator. The cool dry air moves away from the tropics falls at about 30° north and south latitude preventing precipitation, that's why large deserts are located at similar latitudes around the world.

Large mountains are another source of local and regional variation in climate. When you move up a mountain, the average temperature decreases while precipitation increases. These changes are caused when prevailing wind currents encounter a mountain; the air is forced upward. Similar to the tropics, the air will rise and cool causing more precipation. That's why larger mountains will often have forested ecosystems with higher elevation. In Albuquerque, the average precipitation is 12 inches per year, but the Sandias, rising over 5,000 feet above the city, may receive 30-40 inches of precipitation per year. Large mountains can also remove enough water to create much drier conditions behind them. Nowhere is this more evident than in India where air currents hit the Himalaya Mountains, which forces them to drop their moisture. North of the Himalayas, the climate is much drier. In fact, the driest desert in the world is the Atacama Desert in Chile, which lies in the rain shadow of the Andes Mountains.

Disturbance

Disturbance is also very important for determining where species can exists. Similar to climate, disturbance can be complicated and there are different types of ecological disturbances. In general terms, disturbance is a temporary change to the environment associated with the removal of biomass from the ecosystem. The removal of biomass is the loss of organic matter; if a fire burns the fuel on the ground, then there has been a loss of biomass. Most disturbances including fire, flooding, grazing, or high winds may act quickly lasting a few hours to a few days. Other disturbances, including a prolonged drought, may last weeks to several years. Contrary to popular belief, natural disturbances, including fire and flooding, are important for maintaining high levels of diversity as many plants and animals actually depend on disturbances. When natural disturbances are removed from ecosystems, it can be followed by a loss of diversity and a transition to other common species.

Fire: Perhaps one of the most misunderstood types of disturbances by the public is the regular burning of our forests. Fire is an integral part of many ecosystems, including grasslands and most coniferous forests in the west, far north, and the southeast. In these regions, many of the plants are well adapted to frequent fire. For example, Sand Pines (*Pinus clausa*) in the southeast will only open their cones after being heated by fire, meaning new trees can't grow unless there has been fire. Longleaf pines (*Pinus palustris*) require open areas with lots of sunlight for their seeds to germinate. Fire removes other plants and debris, making it easier for the seeds to germinate. Longleaf pines are so dependent on fire that even their needles have evolved to easily burn, but not too hot so that young plants aren't damaged by the fire.

Over a hundred years ago, the southeastern United States from the Carolinas to Texas was dominated by open park-like Longleaf pine forests where ground fires spread through the region. Sometimes the fires would last for weeks burning hundreds of thousands of acres. These fires would occur on a regular frequency of about 3-6 years, preventing the buildup up fuel on the ground. Additionally, the frequent fire would kill other types of trees, preventing the succession to a hard wood forest. As a result, the forests had the appearance of an open parkland with scattered trees and an understory dominated by wiregrass and wildflowers. From the early spring to the late fall, the understory of Longleaf pine forest would be packed with white, yellow, red, and purple flowers. In some cases, the diversity of plants would reach upwards of 50 species per square meter (a little over 9 square feet).

Beginning in the 1920s, these natural occurrences of fire were almost completely suppressed; not just in the southeast, but throughout the western states as well. In the southeast, the loss of fire led to a shift from open forests dominated by Longleaf pines and numerous wildflowers to thick hardwood forest dominated by various species of oak trees. The transition resulted in major losses of plant and animal diversity. In the west, fuel built up on the ground and the density of trees greatly increased leading to more intense catastrophic forest fires that kill all the trees, requiring decades for the forest to return.

The U.S. Forest Service along with other government agencies have been working hard in the last few decades to restore fire as a disturbance to ecosystems by conducting controlled burns to remove fuel and improve diversity. With rampant urban expansion fueled by population growth, many people now live near fire-dependent forests. Unfortunately, many people living near national forests don't like the smoke from the controlled burns, making it difficult to properly manage our forests and restore the ecosystems to a more natural state. The irony is that the lack of controlled fires actually increases the chances of property damage from fire.

Flooding: Similar to fire, river flooding is often considered something bad because of the property damage that results from flooding. To control flooding, various government agencies have built nearly two million water control structures in the U.S. alone. But, like fire, flooding is a periodic disturbance and a natural part of many aquatic systems. Organisms in the streams and adjacent riparian areas have become dependent on regular flooding for their survival. In the Rocky Mountains, spring flooding occurs with some regularity as the snow pack melts and many western streams and rivers will swell and flood their banks. The high water helps remove biomass and replenish underground water supplies. By removing biomass, bare areas are created, which allows many species of plants to seed and grow. The routine flooding can also help prevent any one species from taking over and dominating the landscape, thus preserving diversity.

In New Mexico, along the banks of the Rio Grande River, is a ribbon of forest dominated by cottonwoods (*Populus fremonti*) and called the Bosque (Spanish for forest) by the locals. Historically, the Bosque would be flooded each year by spring runoff from snowmelt at the headwaters located in the San Juan Mountains of southern Colorado. The spring runoff would remove debris, creating bare ground for the cottonwood seeds to land and germinate. Starting the in the 1940s, water control devices including jetty-jacks and dams were put in place to prevent the natural overbank flooding. As a result, cottonwood seeds are unable to germinate to replace older trees as they die. Over time, the large cottonwood trees will be replaced by salt cedar and other non-native trees, further reducing diversity in the Bosque.

Other Abiotic Factors
Light, Nutrients, Salinity, and soil Type

The amount of sunlight and nutrients available to an ecosystem help determine the rate of photosynthesis. This becomes especially important in the oceans where light doesn't penetrate much beyond 700 feet, which is less than 10% of the ocean's average depth. The vast majority of our ocean's environment is in perpetual darkness. Ironically, much of the ocean's surface is a biological desert despite having abundant sunshine and water. These regions are far from any land, which would provide vital nutrients required for phytoplankton growth. In this case, it is nutrients that are limiting where the species are found, not water or energy.

The salinity of water also places limits on where aquatic organisms can live. In fact, the entire phylum Echinodermata is completely limited to marine environments. Not a single species of starfish, sea urchin, or sea cucumber has ever successfully evolved to survive in freshwater. There are more than 30,000 species of fish, approximately half are confined to freshwater environments while the other half are marine. The differences in salinity has resulted in very few fish that are capable of surviving in both marine and freshwater environments.

On land, soil types are also very important for determining where plants can grow. Plants that live in sandy —well drained soils must have the ability to handle periodic dry conditions. The pH of soil also effects nutrient availability to plants. Wetland areas in the southeastern U.S. often have acidic soils that are very poor in nutrients. In these areas, some plants have evolved to be carnivorous, capable of supplementing their nutritional needs by capturing small animals.

Biological Interactions Determine the Realized Niche

The realized niche of a species is where you actually can find it and is determined by biotic interactions. A species' realized niche will always be a subset of the fundamental niche. There are numerous types of biotic interactions including herbivory, predation, mutualism, and parasitism that are important for determining the realized niche of an organism. Biotic interactions can be harmful, neutral, or beneficial, which ecologists describe this as -/-, +/-, +/0, or +/+ relationships. Below, I discuss how biotic interactions determine the realized niche of different species.

Competition (-/-)

I'll never forget exploring a rock outcropping along a remote beach in the Bahamas, it was covered with different species of colorful snails with at least two different sizes. The smaller ones were wedged in tiny cracks in the rocks while the larger ones were found on larger surfaces unable to get into the smaller cracks. Regardless of their sizes, these snails were all grazers, scraping algae off the rocks. Because of their size differences, they were able to coexist on the same rocks while exploiting different areas of the rock. In this example, the larger snail was a type of Nerite, its realized niche that was very similar to its fundamental niche. The smaller snails were confined to small crevices because they could not directly compete against the much larger nerite snails. As a result, the realized niche of the smaller snail species was limited compared to its fundamental niche due to direct competition. By evolving into different size classes, these snails could exist on the same rock by reducing competition between the species, they just utilized different habitats.

Similar types of competition are common throughout the world. If you ever visit a sandy beach, you may notice sandpipers routinely sticking their bill into the sand searching for invertebrates. On a single beach, you may encounter a dozen or more species, each with different body sizes, bill lengths, and habits. Over time, natural selection has driven the evolution of sandpipers into unique species to exploit different resources in the same beach to reduce competition between species, also known as interspecific competition.

Sanderlings and Marbled Godwits easily illustrate how two species of sandpipers divide the resources of each habitat, each occupying a different niche. The Sanderling is a small white bird with a small straight bill often seen running just in front of the waves crashing on the beach. They go after small animals living near the surface of the sand, whereas the much larger Marbled Godwit will use its long bill to probe much deeper into the sand, finding different prey items. These species are dividing the resources of their environment by exploiting different prey items based on bill size. In this case, these birds actually cohabitate on the same beach. Other sandpipers may forage further away from the surf, while others may prefer mudflats in bays and inlets.

Plants also experience lots of competition. Typically, you will find Ponderosa Pines at mid-elevation in the Rocky Mountains. However, they may be able to grow at higher elevations in the mountains. The trade-off for growing at higher elevations is slower growth and its larger branches may be more easily damaged in years with lots of snow. In this case, it is outcompeted by the more narrowly branched fir trees, whose branches accumulate less snow, thus avoiding damage from excess snow accumulation. In this case, the Ponderosa Pine, can live at higher elevations, it's just not found there because it is outcompeted by fir trees which are better adapted to the colder snowier conditions.

These examples of competition between different species illustrate how competition can restrict where a species is actually found. Direct competition between species, costs both species and is therefore considered a -/- interaction. Compeition can reduce the size od a species realized niche. It also illustrates how day-to-day interactions and natural selection act to drive the evolution of species to reduce direct competition. In the case of the snails, competition between species drove the evolution of two very different shell sizes in snails, or different sizes of bills in the shorebirds. This makes sense because direct competition between species is negative for each species. Although similar species may cohabitate in the same areas, they are utilizing the habitat differently based on different traits that were the result of natural selection driving their evolution to reduce competition. As a result, they have different realized niches.

The evolution to reduce competition always involves trade-offs for each species. The larger snails can eat larger pieces of algae and are capable of displacing the smaller species, but their large size prevents them from exploiting the resources found in tiny rock crevasses. Fir trees have smaller branches that prevent them from breaking during heavy snow falls. However, at lower elevations they are outcompeted by the Ponderosa pines with larger branches allowing the pines to gain more access to sunlight. As you can see, there are always trade-offs in evolution, if you evolve to do one thing well (reduce snow accumulation on your branches), you become less well adapted to do something else (growing larger branches to capture more sunlight). Therefore, there are no perfectly adapted species.

Predation (+/-)

The rise of predators at the start of the Cambrian over 542 million years ago may have been one of the major factors behind the evolution of animals. Even today, predation remains a driving force of evolution. Predation is where an animal (or even plant) kills another animal for food. Predation directly benefits the predator (+) at the costs of the prey (-), hence the +/-

relationship. We are most familiar with large charismatic predators like Cheetahs, Lions, Orcas, and Great White Sharks, but the world is full of predators in all shapes and sizes. They have evolved numerous methods of catching prey; Cheetahs, Peregrine Falcons, and Blue Marlin have evolved to be fast to catch their prey. Spiders build intricate webs to catch flying insects. Many species of snakes have evolved specialized venom to kill their prey, while boas and other snakes known as constrictors suffocate their prey. A freshwater fish from South America has evolved to look just like a leaf; it has a very large mouth that it can rapidly open sucking in unsuspecting prey. Even more strange are the electric eels and catfish that have evolved the ability to generate a powerful electrical field so strong they stun their prey to capture it without a fight.

Perhaps some of the strangest looking predators are the deep sea anglerfishes that spend their lives in near to total darkness. The most prominent feature of these predators are their enormous mouths full of sharp teeth, an image straight out of a science fiction horror story. They also possess a modified dorsal fin called an esca that houses bioluminescent bacteria used to attract prey. In the depths of the oceans, an encounter with a potential prey item is rare, so these fish evolved very large mouths with long teeth so that they can grab rather large prey. This way their meal will last them a long time, or at least until their next encounter.

The mode of predation can limit a species habitat preference. For example, Peregrine Falcons are known for their incredible speeds in the air as they dive-bomb medium sized birds such as ducks. However, they are not found in dense forests where they lack maneuverability and would be unable to catch small woodland birds. Specialized predators are going to be found in the same environments as their prey. Likewise, some animals may be limited in their range by predation, if they get pushed to the margins of their habitats, they may be more easily picked off by predators. Small woodland birds may be very agile in flight, but if they flew into the open, they would be relatively easy targets for fast moving falcons.

Symbiosis

Symbiosis means living together, a symbiotic relationship occurs when two or more different species have close interactions, and often for long periods of time. The Disney film *Finding Nemo* popularized the symbiotic relationship of clownfishes with their anemones. Symbiotic relationships can be mutually beneficial when both species benefit (+/+), commensal when one benefits while the other is unaffected (+/0) or parasitic when one species benefits at the cost of the other (+/-). In some cases, the symbiotic relationships are obligatory, meaning that both species must be present for either one to survive.

Mutualism: Ants are an amazing insect; they are found on every continent except Antarctica and they are vital to the functioning of those ecosystems. In the tropics of the Central and South America, leaf cutter ants harvest leaves from the surrounding forest to be used in underground gardens where they grow a special fungus that they harvest. The fungus benefits from the constant supply of food brought in by the ants, and the ants benefit from the nourishment they receive from the fungus. Leaf cutter ants are not the only ants to form mutualistic symbiotic relationships. If you have a flower garden, look closely and you may notice small aphids on the stems of your plants. After a few careful observations, you may also notice small ants tending to the aphids and protecting them from predators. Because the aphids insert their rostrum, a specialized mouth part, into the plant's vascular tissue, they are constantly secreting a sugary liquid which is eaten by the ants. In this case, both species are benefiting from the relationship.

One of my favorite mutualistic relationships occurs in corals, animals related to jellyfish and sea anemones. Coral reefs are home to some of the most productive and diverse areas on the planet. In fact, coral reefs may account to nearly half of all marine species, yet they occur in less than one percent of the oceans. Surprisingly, coral reefs are only found in crystal clear waters devoid of many nutrients resulting in a paradox of how you can have so much productivity and diversity in very nutrient poor areas. Recall from Chapter 9 that the answer lies in the symbiotic relationship between the coral, a small animal, and a small single-celled photosynthetic organism called zooxanthellae that lives inside the coral. The coral sends out its tentacles to catch small prey or debris in the water supplying vital nutrients, while the symbiotic zooxanthellae produces carbohydrates through photosynthesis. Without this mutually beneficial relationship, tropical reefs would not exist and there would be much less diversity in the world.

Parasitism: Many parasites often form symbiotic relationships with their host, or in some cases multiple hosts as they complete their life cycle. Some parasites are even capable of altering the behavior of their hosts to ensure the continuation of their life cycle. One striking example is a liver fluke called *Dicrocelium dendrticum,* which infects cows. Inside the liver, it mates and lays eggs that are excreted in the cow's feces. Small snails ingest the eggs that hatch and reproduce asexually inside the snail. However, the snail will get rid of the parasite by excreting them in a ball of mucus. At this point, ants will eat the snail's mucus and ingest the flukes that will then move to the head of the ant. Once there, they alter the ant's behavior to make them crawl up a blade of grass in the evening to get eaten by a cow to repeat their life cycle. If the ant doesn't get eaten, the fluke will let the ant carry on with its normal buisiness until the next evening.

History Matters

Have you ever wondered why Australia is full of unusual marsupials, such as koala bears and kangaroos? It turns out that when it comes to the distribution of animals such as marsupials, geological history matters. The preponderance of marsupials in Australia is not so much of what happened a few hundred years ago, but what happened hundreds of millions of years ago and continues today. The process of continental drift, the slow movement of continents across the planet over tens of millions of years, is largely responsible for Australia's marsupials.

About 250 million years ago, all the continents were merged into one giant supercontinent called Pangea. Slowly, over millions of years, the super continent broke up into smaller continents that we know today, including Antarctica, South America, North America, Africa, Europe, Asia, and Australia. Mammals first appeared nearly 200 million years ago during the Mesozoic Era, the earliest ones resembled the platypus, which are egg layers. Later, the marsupials evolved live birth and spread across the continents. Marsupials migrated from South America to Antarctica and then to Australia. When Antarctica and Australia broke away from the other continents it isolated the ancient egg laying mammals and marsupials from modern mammals. As a result, the marsupials were able to diversify into unique lineages including kangaroos and koala bears.

New Zealand is a continental fragment, it's not quite a continent, but larger than an island. It broke away from Australia and Pangea before any mammals had evolved. As a result, it has existed for over two hundred million years without mammals, except for bats that arrived by flying and some invasive rodents brought by man. Similar to Australia, New Zealand is home to ancient lineages of birds like the Kiwi and a unique lizard-like reptile called the Tuatara. It superficially resembles a lizard, but has been separated from lizards for over 220 million years ago. Tuataras are a living fossil, replaced on other continents by modern lizards. The long separation of New Zealand from other continents has allowed this ancient lineage of reptiles to exist for over 200 million years.

Madagascar is another continental fragment that broke away from Africa nearly 60 million years ago, carrying with it a lineage of mammals called primates. Humans along with monkeys and great apes are in a group of primates called the Simians. On Madagascar, there are approximately 100 species of lemurs, an ancient lineage of primates called Prosimians that still survives today because they have remained isolated from other modern species of primates. Unfortunately, many are threatened or endangered due to habitat loss and hunting pressures.

For the past 2.5 million years (although some geologists argue we have been in an ice age for nearly 35 million years), the Earth has been in an ice age, a period of long-term cooler temperatures, continental ice sheets, and permanent ice at the poles. In fact, we are still in an ice age, just not a period of maximum glaciation. During the last glacial maximum that ended a mere 11,700 years ago, large glaciers covered vast areas North America and Europe. So much water was locked up in the ice sheets that the sea level was approximately 300 feet lower than it is today. In North America, glaciers covered much of Canada and extended southward to Kentucky. When they receded, plants and animals colonized the bare lands left behind. Small isolated pockets in the Appalachian Mountains that remained glacier free formed refuges for a number of species that would spread after the retreat of the glaciers.

During the last glaciation, many islands were connected, but are separated by water today. In the 1850s, a British Naturalist named Alfred Russel Wallace was traveling throughout the Indo-Pacific collecting animals and plants for museum collections. He noticied that Australia, Tasmania, and New Guinea shared similar species, but there was a big change in species composition between New Guinea and Borneo. He also made similar observations on several islands where some were more similar to Southeast Asia. Now we know that the difference in species is the result of historical connections between New Guinea and Australia during the last ice age, or between islands and Southeast Asia. We call the line between the two regions the Wallace line. During the last ice age, the two regions remained separated by a deep water channel preventing the mixing of Australian and Southeast Asian species.

What Limits Population Growth?

For the first part of this chapter, I focused on the factors that affect the distribution of species, but what factors limits the growth of populations? A population is a group of individuals of the same species. The easy answer to this question is that birth rates and immigration will add to a population while death rates and emigration will reduce population size.

The more complex answer lies in what determines death and birth rates or emigration and immigration rates. Over two hundred years ago, the English scholar Robert Malthus understood that any population is capable of exponential population growth. It goes something like this, imagine a single *E. coli* bacterium that divides into two daughter cells, then each daughter divides and there are four cells, then eight, then 16, 32, 64, 128, 256, 512, 1000, 2000, 4000, 8000, 16,000, 32,000 and so on. Within a few days, the Earth would be covered in *E. coli*. Exponential population growth is not limited to rapidly dividing bacteria. If you were to take a single pair of houseflies at the start of summer, if all their offspring and all their offspring's offspring (grandflies) were to survive throughout the summer, there would be more than 325 trillion individuals This would be enough houseflies to circle the Earth's equator 57,000 times! Luckily for us, we are not covered in houseflies or *E. coli*. What are the drivers of birth and death rate, or immigration and emigration?

Ecosystems have a maximum amount of resources available, there is only so much energy, nutrients, and space available. Taken together, the available resources determine an ecosystems carrying capacity, which is the maximum population size that can be reached. In many areas, carrying capacity can fluctuate based on climate patterns in the area.

Populations of plants and animals will respond to changes in the carrying capacity as environmental conditions fluctuate. When resources are abundant, the carrying capacity of an ecosystem will increase and populations will grow. Sometimes, a population can grow too quickly, overshooting the carrying capacity of the ecosystem. When this happens, the population will decline. Likewise, if a resources becomes scarce, as happens during drought, then populations respond by becoming smaller as individuals either die or immigrate to other areas.

Deserts provide a good example to illustrate how carrying capacity can fluctuate. Deserts receive plenty of energy from sunlight; however, their carrying capacity is limited by water. In the Southwestern U.S., deserts depend on monsoonal rains beginning in July. In wet years, the rains lead to more plant growth, which in turn provides more food for herbivores of all types from insects to antelopes. When populations of insects drastically increase in response to more water and food, this will also increase the populations of insect eating birds in the area. Therefore, in wet years, the carrying capacity for the desert is higher. In dry years, plants respond by growing much less and producing fewer seeds. As a result, there are fewer herbivorous insects leading to fewer insect eating birds. In parts of New Mexico between 2010 and 2014, Burrowing Owl populations drastically declined, a direct result of prolonged drought in the region reducing the area's carrying capacity for this species. The drought drastically reduced plant growth, which in turn caused a decline in rodent populations the main source of food for Burrowing Owls.

The ability of a species to respond to changes in the carrying capacity can vary based on the length of their lifecycle, and number of offspring produced during a given time. Insect populations may fluctuate wildly from year to year due to their rapid generation times. Black Bear, on the other hand, have much longer life spans and lower reproductive rates and will respond much more slowly to changes in carrying capacity. If their populations were to become severely reduced, it may take them several years to recover.

How quickly a species responds to fluctuations in carrying capacity depends in large part on its survivorship curve. A survivorship curve is used to predict the proportion of individuals surviving to a certain age class. In general, there are three different survivorship curves, Type 1, Type 2, and Type 3. Humans and whales exhibit a Type 1 curve; we have a high probability of making it from one year to the next. Over 50% of people born will make it to the average maximum age of the species. As individuals begin to approach the maximum age, the rate of mortality from one year to the next drastically increases. In a Type II curve, individuals have about the same chance of surviving from one year to the next. For example, a squirrel may live a maximum of about 5 years, but each year, it has a 20% probability of death. Animals with a Type III curve experience the greatest mortality early in life. Spadefoot toads may lay hundreds of eggs, but almost all the tadpoles and young frogs are likely to die in the first year. However, once they make it to adulthood, they have a much higher likelihood of surviving a number of years.

The Day to Day Ecological Interactions Drive Adaptive Evolution

Evolution, the change in a species over time, is a fact because it is something that actually happens or exists. We can observe it, measure it, and experimentally verify that it is taking place. Why species change over time is best explained by Darwin's theory of evolution by natural selection. Recall that natural selection is based on the assumption that there is variation within a population. Some individuals are more fit to the environment, and as a result they are more likely to survive, reproduce and pass on their favorable genes to the next generation. Over time, favorable traits accumulate in the population and it changes over time. Because species change over time, we are all descended from common ancestors.

The rate of evolution can be quite rapid as we see in the evolution of antibiotic resistant bacteria to imperceptibly slow as seen with the horseshoe crabs, whose basic form has not changed in nearly 450 million years! Although modern horseshoe crabs may look similar to their ancient ancestors, they are still undergoing evolution. It's much a like a Volkswagen car from the 2000s looks superficially similar to one built in the 1960s, but basically every aspect of the electronics and engines have changed to become more efficient.

What does it mean to be better fit to the environment? It comes down to the day to day survival of finding food, avoiding predation, fighting off diseases and parasites, and reproducing. If the climate you find yourself in becomes too hot, do you have ways to remain cool, like sweating or panting? Can you search for shade or bury yourself underground? If it becomes dry, can you conserve water, or easily move to new water sources? If predators are present, can you outrun them, avoid detection, or fight them off? If you are a group of male prairie chickens on a lek, is your song and dance enticing enough for the females? Are your feathers bright enough to attract the females? All of these factors are important for survival and reproduction.

As you can see, the day to day ecological interactions place selective pressures on individual organisms. Surviving a cold spell, evading a predator, finding food, and reproducing occur on the short time scales of hours, days, even weeks. These selective pressures act on the individual, either you can survive and reproduce or you don't. In the end, an individual does not evolve, it is the populations that evolve over time. Sometimes, populations cannot adapt quickly enough to rapidly changing environments and they perish. If all the populations of a species are lost, then it is extinct. It has been estimated that 99% of all species that have ever lived on the Earth have gone extinct. That means that the roughly two million species alive today represent only 1% of all the life that has existed on the Earth!

Every adaptation evolved in a species comes at a cost, that cost is the loss of being able to do something else well. For example, a cactus can survive in extremely dry environments due to multiple adaptations. It has shallow roots to quickly absorb water from infrequent rains and their leaves have been reduced to thorns serving a two-fold purpose of protecting the plant from herbivores and to prevent water loss. Cactus further reduce their water loss by opening their stomata only at night or early in the morning when the humidity is highest for the day. Stomata are tiny openings that allow gas exchange with the atmosphere and are important for the movement of water from the roots.

These adaptations help cactus survive in arid environments where few plants can, but the trade-off is that they grow very slowly. In fact, the saguaro cactus of the Sonora desert may live for hundreds of years. Even if you were to provide it with plenty of water, it still wouldn't grow much faster. Ironically, if you water a cactus too much, it will actually die because it is not adapted to too much water. Cheetahs may be the fastest land animal, but that also limits them to much smaller prey than a lion, which weighs almost 4 times more than a cheetah. Cheetahs would not survive well in a forest because trees would prevent them from moving at full speed, allowing their prey to more easily get away. Evolutionary trade-offs to survive in an environment means that there are no perfectly adapted organisms. You can't be good at everything at once.

A Saguaro Cactus in the Kofa National Wildlife Refuge in Arizona. Cactus are well adapted to living in desert environments by preventing water loss. However, the adaptations of cacti in deserts is a trade-off, if you supply them with lots of water, they will not grow as fast as other plants adapted to wetter environments.

Chapter 12

Ecosystems Connect
Life to the Earth

*When you get right down to it,
life is an extension of geological cycles*

John Rogers

Introduction

The amount of energy and nutrient availability influences a vital component of ecosystems called primary productivity. In most ecosystems, primary productivity is tied rates of photosynthesis, which can fluctuate over time. In turn, higher rates of primary productivity are usually correlated to higher rates of diversity.

Regardless of the type of ecosystem, they must all recycle materials, including carbon, nitrogen, phosphorous, and water. Life is connected to the Earth as nutrients constantly move back and forth between the environment and living organisms. Most ecosystems rely on plants and other photosynethetic organisms to harness the energy in sunlight to make these materials available to life. As the energy is transferred through different trophic levels in an ecosystem, the energy that entered as sunlight will eventually exit the ecosystem as heat.

You may not notice it, but ecosystems are dynamic and changing over time. Sometimes the changes can be rapid, while at other times, the changes can be incredibly slow. Not only do ecosystems change in time, they are also connected to each other. Often the narrow areas where two different ecosystems connect can harbor unique species and increasing the diversity across the landscape

Similar environmental conditions, such as a wet and warm climate, leads to similar ecosystems even though the actual species comprising them can be quite different. Similarities in ecosystems let us categorize them into the biomes. You may already be familiar with several biomes including deserts, coniferous forests, temperate broadleaf forests, or grasslands.

Living Organisms Form Communities

If you have ever taken a hike on a trail surrounded by tall evergreen trees, or hardy desert vegetation, then you are in a biological community. All the species living and interacting with each other in a certain area forms a community. A community can be small, such as the microbes living inside your gut. Or, a community could be much larger, such as all the fish and invertebrates living in a nearby lake.

Biological communities are defined by the types of species present and their interactions with each other through competition, predation, parasitism, and other symbiotic relationships. In the last chapter you learned how these interactions were influential in determining the realized niche of a species, or where you actually find a particular species. But, all these species must also exist in their physical environment, which determines their fundamental niche, or where they can actually be found based on physiological limitations. Therefore, an ecosystem includes the living and non-living components in an area interacting with each other.

When traveling from one region to another, whether it is driving up a mountain, or driving to the coast, you will notice diversity varies across the landscape. The types of species change based in large part on the climate and other aspects of the environment. At the base the Rocky Mountains you may encounter mostly pinyon pines and juniper, while at higher elevations, you may encounter slender fir trees adapted to the colder environment with more snow. In marine environments off the west coast of North America, you may swim from areas with hard rocky bottoms dominated by sea urchins grazing on encrusting algae and then abruptly enter into a kelp forest populated with thousands animals. Whether it is a marine or terrestrial community, they can vary in the number of species present, the relative abundance of the species, and the types of species. Communities are dynamic systems, changing over time as some species come and go, or change in their relative abundances.

The number of species present in a community is often referred to as biodiversity or simply, diversity. Diversity can vary immensely between communities in different regions. For example, a tropical rainforest may have 300 species of trees per acre, whereas a northern boreal forest may have less than 10 species of tree per acre. Differences in diversity in different regions, such as a tropical rainforest and a boreal forest, are largely due to the climate and other aspects of the physical environment. In the next sections I will explain why rain forests are more diverse than a temperate forest, or a desert.

The relative abundances of species within a community can also vary. There may be upwards of 300 species of trees in an acre of tropical rainforest. In these tropical communities, each species may be represented in relatively equal proportions. Whereas, in Longleaf pine forests, the pines have a very high relative abundance where 90% or more of the trees in an acre are Longleaf pines. Temperate deciduous forests may have higher tree diversity than Longleaf Pine forests, but 50% of the trees in an area could be comprised of just one or two dominant species.

For those of you interested in natural history and are familiar with the plants and animals in your area, you know that most species are rare. Most of us are familiar with Rock Pigeons, American Robins, and House Finches, which are common birds in most towns. While there may be 50 House Finches in your neighborhood, there may be only a pair of Greater Roadrunners, or a single Cooper's Hawk. The answer to why a few species are common while most are rare is complicated. Once again, both the physical environment and species interactions are important for explaining this general pattern. Predators are rare simply because only about 10% of the energy is transferred from one trophic level to another. Sometimes, the presence of a certain species can have a very large impact on community structure, determining who is else capable of living there.

I Return to the kelp forest to provide an example of the importance of a single species to an entire community. The presence of the kelp provides structural habitat for numerous fish and invertebrates leading to very high rates of diversity. This entire community is dependent on the presence of just one species, the Sea Otter, a keystone species. A keystone species does not have to be the most abundant animal in the community, but they do have a large impact relative to their population. Sea otters are not the most abundant species inhabiting the kelp forest, but they have an enormous impact on it. Sea Otters are the natural predators of sea urchins, which are voracious herbivores of kelp. When Sea Otters are removed from kelp forest, the urchins begin to proliferate and eat the kelp at its base. Over time the urchins will eventually cause the complete loss of the kelp forest and the hundreds of species dependent on it.

There are other examples of keystone species. Beavers build dams on streams causing the water to pool behind their dams creating natural wetland areas. These wetland areas provide habitat for numerous species of plants and animals requiring the wet habitat. The Red-cockaded Woodpecker is found in mature Longleaf pine forests where it drills its nesting cavities into live pine trees. The abandoned cavities are used by Eastern Bluebirds, Carolina Chickadees, Brown-headed Nuthatches, flycatchers, other woodpeckers, Eastern Screech-Owls, and even Flying Squirrels. Without the Red-cockaded Woodpecker present to make cavities, these species would decline in population from lack of suitable nesting habitats.

All Ecosystems Cycle Nutrients

Life is connected to the Earth as nutrients continually cycle between us and the environment. It's a process that has been ongoing for billions of years starting when life began. The first living organisms were most likely autotrophs that used naturally occurring sources of energy to fix inorganic carbon into organic forms. Upon the evolution of the first living organisms, nutrient cycling was forever changed. Carbon, for example, could move quickly between the atmosphere and living organisms, or it could be sequestered as rock for millions of years as part of longer geological cycles.

All life requires nutrients to exists, they can be either organic or inorganic. Organic nutrients are organic molecules that includes carbohydrates, proteins, lipids, and vitamins. We can use organic nutrients as a source of energy and building blocks to make other organic molecules required for our functioning. Most of these organic molecules are eventually broken down to carbon dioxide and water. Inorganic nutrients include small inorganic molecules such as water, carbon dioxide, nitrates, ammonia, phosphates, along with variou minerals and electrolytes including iron, calcium, potassium, chlorine, sodium, and *etc*. Carbon dioxide is considered an inorganic molecule because it is not bound to any hydrogen atoms. As an animal, we obtain all of our nutrients from the foods we eat and the water we drink.

Nutrients must be continuously recycled because they are not being destroyed or created. These nutrients go back and forth between an inorganic reservoir and living organisms. For example, the atmosphere serves as reservoir for both carbon and nitrogen, and the oceans are the major reservoir for water. While there are many types of nutrient cycles, I'm only going to discuss carbon and water.

I'll start with carbon to show how nutrient cycles connect us to the Earth. You have more than a trillion, trillion carbon atoms in your body, each one is more than 5 billion years old. Additionally, every carbon atom in your body was once in the atmosphere as a molecule of carbon dioxide, perhaps as recently as a few weeks ago. Some of those carbon atoms may have once been in a rock 50 miles beneath the surface in the Earth's mantle, or in a piece of coal, or once formed a protein in a dinosaur a hundred million years ago. When the dinosaur died, its carbon would have been released into the atmosphere as carbon dioxide. Once in the atmosphere, that carbon atom could have been fixed into a carbohydrate through photosynthesis. Later, an herbivore would come along eating the plant and acquiring that carbon atom, only to breathe it out again. And so the carbon cycle continues, carbon moves between the living and the non-living world as it is continually recycled.

A carbon atom may be in a molecule of carbon dioxide in the atmosphere or part of a complex organic molecule inside of a cell. The metabolic processes of aerobic respiration and photosynthesis constantly cycle carbon between inorganic carbon dioxide and organic carbon. The last type of chemical reaction will determine what type of molecule carbon may be a part of. No matter how many times carbon is cycled through the environment, or how many chemical reactions it has been subjected to, it is still carbon. That's why you may have carbon atoms in you that were once in a dinosaur or a rock.

The element carbon is not only important for organic molecules. As carbon dioxide it is an important greenhouse gas because it is able to trap heat. Therefore, when CO_2 levels are high, the climate tends to be warmer. Life has played a crucial role in moderating atmospheric carbon dioxide levels by balancing its removal with inputs from natural sources, such as volcanism. Carbon dioxide can also be removed from the atmosphere when it dissolves into the oceans forming bicarbonate. Small marine organisms will then form shells of calcium carbonate, a type of rock that they use for protection. When they die, their carbon atoms exit the quick back and forth movement between living organisms and the atmosphere. Instead, their carbon enters long-term geological cycling taking place over millions of years as it forms carbonate rock known as limestone. Eventually, those rocks may be buried or even subducted into the interior of the Earth where the immense heat and pressure will form carbon dioxide that will be released in volcanic eruptions.

In the last 400 million years, life has also removed large amounts of carbon from the atmosphere by forming coal and oil deposits. In fact, our rapid burning of fossil fuels is releasing carbon into the atmosphere that has been out of circulation for hundreds of millions of years. Once released back into the atmosphere, it is once again able to be incorporated into another organism when it gets fixed by plant. And so the carbon atom is cycled over and over again. It continuously moves between plants, animals, rocks, and the atmosphere.

One of the most vital components to life is water. Water is an essential molecule for life, it is a good solvent providing the medium for all the chemical reactions of life. Unlike carbon, water is a molecule that can be made and broken down through chemical reactions. In your body, there may be water molecules hundreds of millions of years old. Similar to carbon, you may have some water molecules in you that were once in a dinosaur. Other water molecules in your body may be only seconds old, created by metabolic reactions taking place in your cells.

Generally, when ecologists discuss water cycles, they aren't too concerned about water being broken down and created through chemical reactions (remember the oxygen and hydrogen atoms that form water are not altered by chemical reactions). Our oceans form an incredibly large and permanent reservoir of water. Over time, water moves from the oceans to the atmosphere, to the land, through organisms, eventually making its way back to the ocean. The vast majority of water molecules will repeatedly make this journey, sometimes lasting only a few days. Other times, water may seep deep into the ground taking thousands, or even millions of years to return to the oceans. When you drink a glass of water, think of the journey those water molecules have been on. If your water came from ground water, they could have been there for decades to millions of years depending on the aquifer. If your water came from a river, then some of those water molecules may have been in the ocean only a few days ago.

One of the interesting implications of nutrient cycling is that the elements and molecules in our body continuously move between living organisms and the environment. As I discussed earlier, the carbon atoms in your body could have once been in another living organism, or they could have just as easily been part of a rock, or even pumped from the ground and used as gasoline in someone's car. It makes sense to think of life as an extension of geological cycles. Pick up a rock, the elements that form the rock are the same age as the elements in your body, they have been cycled through geological processes just like the carbon, nitrogen, and oxygen atoms in your body. In fact, it's easy to think of life as an extension of geological processes that have been taking place for billions of years.

Energy Flows through Ecosystems

In *The Empire Strikes Back*, the ancient Jedi master Yoda tells Luke that the force flows through life. If he were talking about energy, he would be right. Ecosystems and living organisms are made possible and powered by a constant flow of energy. The ultimate source of energy for almost every ecosystem is sunlight, and most of the energy that enters an ecosystem will eventually exit as heat. Energy is unlike nutrients; it cannot be recycled. Recall that pesky second law of thermodynamics, it tells us that every time energy is used, some of it is converted to heat, a relatively unusable form of energy.

At the base of every ecosystem are autotrophs that use energy to make inorganic carbon biologically available to other living organisms. Autotrophs form a link between geological or atmospheric cycles and living organisms as they bring carbon and minerals into food webs by forming organic molecules. The first autotrophs were likely primitive prokaryotes, adding hydrogen to carbon dioxide by harnessing natural energy sources in ancient alkaline vents. In the deep sea, where most thermal vents are found there is no light. In these ecosystems, autotrophs do not rely sunlight, they extract energy from rocks, minerals, or hydrogen gas. However, all that changed sometime about 3.5 billion years ago when ancient cyanobacteria evolved the ability to use the energy in sunlight to fix carbon dioxide into organic molecules with the benefit of storing energy for later use.

Heterotrophs depend on autotrophs for all their energy and nutrients. Heterotrophs are quite varied in terms of how they acquire energy and food from their environment. In broad terms, we label heterotrophs as decomposers, or primary, secondary, or tertiary consumers based on how high up the food chain they are. Herbivores would be primary consumers, eating only plant material. Carnivores would be secondary or tertiary consumers eating herbivores or even other carnivores. Carnivores may not eat plant material, but their energy and nutrients still began with plants. Humans are a great example of omnivores, animals that eat both plant and animal material.

On average, when one organism eats another organism, whether it's a plant or animal, only about 10% of the energy is transferred, the rest is lost to the environment as heat. Although, the amount of energy transferred can vary depending on the organism. Endothermic animals, such as birds and mammals, spend a large amount of energy generating heat to remain constantly active. Therefore, only about 1-3% of the energy they acquire is used for growth or stored, that's why we have to eat so much. Ectotherms, such as a snake, may have a metabolic rate 100 times less than a human when corrected for size differences. That's why they can survive for a month on a single meal, whereas we would starve to death in a few days.

Another way to think about the energy flow from plants to herbivores to carnivores, is to imagine a patch of grass with about 10,000 Kcal of energy, only about 1,000 Kcal of energy would be transferred to herbivores, and 100 kcal of energy would be transferred to carnivores. Based on this energy transfer, it becomes apparent why the world is green and herbivores are rare, and predators are even more scarce. The largest animals on the Earth are able to maintain their enormous size by going to the base of the food chain to feed mostly on krill, a small crustacean that feeds on phytoplankton.

Decomposers are heterotophs that break down organic matter. They oxidize carbon back to carbon dioxide and they return inorganic nutrients, such as nitrogen, phosphorous, calcium and potassium, back to the ecosystem so they can be absorbed once again by plants. While fungi are well-known decomposers, the vast majority of decomposers are actually bacteria who largely go unseen in ecosystems. It's been estimated that the biomass of all the bacteria on the planet far exceeds the larger and more easily observed eukaryotes, such as plants and animals.

Autotrophs form a direct link between geology and biology as they use energy to make nutrients available to ecosystems. For example, photosynthesis requires energy to make carbon available to living organisms and nitrogen fixing bacteria require energy to make nitrogen available to some plants. Without energy, carbon would quickly become locked up in carbon dioxide and inorganic rocks. Nitrogen would return to the atmosphere as nitrogen gas (N_2), which is just as unusable to most life as carbon dioxide. Other elements and minerals would also quickly become unavailable to life as well. Without energy, life would quickly perish as chemical cycling would grind to a halt. Eventually the Earth would head towards a stable equilibrium where no living organism could exist.

When plants use energy to fix carbon dioxide, they are moving carbon into an ecosystem by adding biomass or organic material. One of the most important ways to measure an ecosystem's functioning is by how much new biomass is added in one year per unit area. We measure the rate biomass is added to an ecosystem as primary productivity. This is the amount of organic mass added per unit area over a specific time. To understand primary productivity, imagine your lawn: during the summer your grass grows, if you grow one kilogram of grass in a square meter over a year then your net primary productivity would be $1kg/m^2$ -Yr.

If you've ever owned a plant or a grass lawn, then you know that plant growth is positively correlated with how much water you give it. The more water available, the faster your plant grows, and the higher its primary productivity. Based on this simple observation, we know that primary productivity is directly related to process called evapotranspiration, which is the amount of water emitted from plants. High rates of evapotranspiration result from plenty of water availability, abundant sunshine, and nutrients required by plants to grow.

In general, the primary productivity of terrestrial ecosystems is greatly affected by moisture and energy availability. The amount of energy available to ecosystems varies across the planet. Areas near the equator continually receive abundant energy from the sun throughout the year and have the highest rates of primary productivity. As you move north or southward, there is increasing seasonal variation in sunlight due to the tilt of the Earth on its axis as it revolves around the sun. This tilt makes sunlight come in at an angle at higher latitudes and therefore has to pass through more of the atmosphere. So even when it is daylight in winter, the sunlight doesn't deliver the same amount of energy as the direct sunlight at the equator. We witness this every day, it is hottest at noon when the sun is overhead and cooler in the evening and mornings when the sun is low on the horizon. As a result, northern and southern latitudes, on average, have lower primary productivity simply because there is less energy available, especially in the winter when temperatures fall below freezing in large regions of the north.

Terrestrial ecosystems can also vary greatly in productivity based on water availability. Deserts can be quite hot with plenty of sunshine, but water is scarce. As a result, primary productivity is very low. Worldwide, ecosystems with the highest annual primary productivity are also among the most biologically diverse, whereas those with the lowest primary productivity are the least diverse. This phenomenon is easily observed when you map patterns of diversity against net primary productivity.

The interplay between nutrient and water availability along with energy inputs determine the overall productivity and diversity of ecosystems. This goes for both terrestrial and aquatic ecosystems. Although, aquatic ecosystems are often very sensitive to changes in nutrient levels, especially nitrogen and phosphorous concentrations. In freshwater and marine ecosystems, clear water is associated with low levels of nitrogen and phosphorous, which limit the abundance of photosynthetic organisms. When nutrient levels rise, many types of algae and phytoplankton are able to quickly use the resource, greatly expanding their numbers, reducing visibility and light penetration.

When there are too many nutrients, especially nitrogen and phosphorous, it can lead to an unhealthy condition called eutrophication, which causes algae blooms. We often associate higher rates of productivity in an ecosystem with higher diversity. Unfortunately, this does not always hold true with aquatic ecosystems. Increased algae growth can actually reduce plant growth, thus degrading aquatic habitats for many fish and invertebrates. The real problem occurs when the phytoplankton die. Because oxygen is used to break down their remains, eutrophication can cause oxygen levels to drop to the point where animals can no longer survive. A prime example is where the Mississippi River empties into the

Gulf of Mexico bringing tons of nutrients to the region causing a dead zone nearly 6,500 square miles in 2015. Low oxygen levels can also occur at night when photosynthesis stops producing oxygen and respiration becomes dominate, reducing oxygen to very low levels. When nutrient levels rise on coral reefs, algae can grow directly on top of the corals and kill them.

Ecosystems are Connected
Across the Landscape

At first glance, it may seem that some ecosystems are isolated from each other, especially aquatic and terrestrial ecosystems. The water's edge forms a natural barrier for many organisms, thus separating terrestrial ecosystems from aquatic ecosystems. The reality is that ecosystems, even aquatic and terrestrial ones, are intricately connected across the landscape. When two different ecosystems meet, the area of transition is called an ecotone. A prime example is a beach where an aquatic environment connects to the terrestrial environment. Ecotones can also form when between adjacent terrestrial ecosystems too, such as a field and a forest, or when alpine forests quickly transition to alpine tundra.

A consequence of ecotones is that they often have unique assemblages of plant and animal communities; in turn, increasing the overall diversity across the landscape. You may have seen this in your travels. If you've ever visited a beach you may have noticed numerous small shorebirds foraging on the algae, plants, and animals washed up on shore. If you have walked along a river bank, you may have noticed numerous spiders or predatory insects hanging out near the water's edge. You may not find any of these species outside of their specific ecotones.

Ecotones between aquatic and terrestrial environments can also be dynamic, undergoing regular seasonal variation. In northern regions, many rivers have higher flows in the spring and lower flows in the summer, a result of spring snow melt at their head waters. When rivers are flowing at their highest either from routine spring flooding or from heavy rainfalls, the rivers will swell and flood their adjacent wetland areas. The fast moving water can alter the river channel and create new habitats. As the adjacent wetlands become flooded, leaf litter from trees lining the rivers provide nutrients and energy to the river. Later in the summer, the waters recede with reduced flows causing algae production to ramp up due to the clearer water.

The connectivity between two different ecosystems can easily be seen during the summer. Each year as algae production ramps up in the summer it provides food for larval insects that have aquatic stages. Once the insects become adults, they will emerge from the water to carry out their life cycles on land. Terrestrial predators, including birds, lizards, insects and spiders can be found in the ecotone preying on these emerging insects. These predators may account for 50% to 75% of the biomass in this ecotone, far above the typical 1% to 10% you would find in most other ecosystems.

The reason why these ecotones between aquatic and terrestrial ecosystems are diverse with large populations of predators is because they are being subsidized by an input of resources from the aquatic ecosystems. The insects emerging from the rivers relied on algae production as larvae. When they emerge as adults, they provide an input of energy and nutrients directly to the shoreline, where they are preyed upon by birds, reptiles, and other arthropods. Likewise, in the fall, deciduous trees drop their leaves into aquatic communities adding additional sources of nutrients to larval insects including mayflies, stoneflies, and caddisflies, which in turn can be eaten by fish. When these insects emerge as adults, they are consumed by terrestrial predators. As you can see based on this example, terrestrial and aquatic ecosystems are intricately connected through the flow of nutrients and energy.

I like to take long walks on the beach and poke at dead things with a stick, you never know what you may find, or what may move. There many different types of ecotones where marine and terrestrial environments meet. These include, white sandy beaches, mangroves, estuaries, and mangrove forests, each one harboring unique communities. I personally enjoy white sandy beaches because sea grass routinely washes up on the shore providing an input of nutrients and energy to the beach. A quick inspection of washed seaweed will reveal numerous species of small, mollusks, crustaceans, and assortments of small worms, which in turn, are eaten by numerous shore birds and insects. Therefore, a sandy beach with almost no primary productivity from plant growth can host a diverse community of invertebrates and shorebirds by relying on marine inputs.

Not everything that arrives on a beach from the ocean is dead. On the east coast of North America and the Gulf Coast during the spring high tides, ancient looking horseshoe crabs crawl up to the edge of the surf by the tens of thousands to lay their eggs. Their egg laying event is so predictable that some species of shorebirds actually time their spring migration to take advantage of easy meals from all the eggs being laid.

Mangrove forests are another unique habitat found in tropical regions on the world. In the New World, there are three species of unrelated trees called mangroves. They are able to grow from the land and into the water because of unique adaptations, especially in their roots. Red mangroves have unique roots that help prop the tree above the water. Their roots also provide great habitats for many species of fish, and the roots themselves provide a place for algae, sponges, mollusks and tunicates to grow.

Ecosystems are Dynamic

Ecosystems are dynamic and constantly changing over time. Changes can be rapid, slow, predictable, or unpredictable. Any change in climate, alterations in disturbance regimes, such as fire and flooding, or the spread of invasive species can all cause ecosystems to change. Ecosystems also respond to predictable seasonal changes as well.

Seasonal variations between winter and summer in temperate ecosystems are very predictable as they undergo yearly changes in primary productivity. During the winter months, deciduous trees lose their leaves to avoid freezing temperatures and primary productivity is drastically reduced. Once it warms again in the spring, the trees once again grow their leaves and plants begin to photosynthesize and grow. The seasonal changes in primary productivity is enormous, ranging from nearly zero in the winter to rivaling a tropical rainforest.

Ecosystems can also change over longer periods of time. After the last ice age, the glaciers retreated across North America leaving mostly barren ground. Slowly, a process called succession took place where the ecosystems transitioned from small low lying plants, to small bushes, and eventually to forests with tall trees. Succession is where there is a change in the community composition over time. Old fields that are no longer planted can also undergo succession eventually forming a forest. When new oceanic islands form they will also undergo succession as plants and animals slowly colonize the new habitat. On islands, the colonizing species usually come from the nearest land.

Changes in climate and disturbance regimes can also cause ecosystems to change. Some of these changes can be natural, taking place over millions of years as the continents wander across the Earth, slowly causing climate change. Millions of years ago, Antarctica was forested, but as It drifted over the South Pole, it cooled and eventually froze over. Today, the continent moslty covered in ice, and there are no forests present anywhere.

Currently, we are altering disturbance regimes indirectly through climate change, and directly by specific activities. Climate change is already causing communities to change worldwide. Nowhere is this more evident than in the southwest from New Mexico to California where the region is experiencing a prolonged multi-year drought. In New Mexico, the foothill ecosystems near the Rocky Mountains were once dominated by pinon pine and juniper trees. After multiple years of intense drought from the first part of this century, the vast majority of the pinon pines in this area died, leaving mostly juniper trees. In the mountains of southern Colorado, fir trees and aspen trees are dying from a combination of drought and bark-beetle infestations.

In California, where they have also experienced severe long-term drought, similar changes are taking place where large trees are dying, changing the composition of the communities from wooded forest to more scrublands. If adequate precipitation does not return, the average amount of forest will be reduced and replaced by more drought tolerant shrubland and grassland communities.

Routine fire is a natural disturbance in many forests where it removes biomass and recycles nutrients. However, climate change is likely the reason for continuous drought and higher temperatures, which has caused intense forest fires, killing large stands of trees in a few days. The return of these forests is dependent on adequate precipitation in the region. If precipitation remains low, then forests ecosystems will not return and hardier drought resistant shrub-like vegetation or grasslands will replace the forests.

Changes in natural fire regimes are often caused by us. In the southeast where fire has been excluded for decades, Longleaf pine ecosystems are being replaced by less diverse hardwood communities. The succession from pines to hardwoods has been slow, taking decades. But, throughout the southeast, large remnant Longleaf pines, hundreds of years old, are found mixed in with much younger hardwoods, such as oaks and hickories. Unfortunately, these large pines are unable to replace themselves as the young pines require plentiful sunshine and do not grow well in the dense understory of a hardwood hammock. What most people don't realize is that the succession to a hardwood forest actually reduces diversity in the region. Also, the lack of fire in the western forests has actually led to more trees per acre. While this may sound like a good thing, it actually lowers diversity by preventing wildflowers from blooming. Also, when fire does occur, it is more likely to burn hotter and kill more trees.

Most rivers experience seasonal variation in water flow that the organisms in the river and the adjacent riparian areas have become dependent upon to complete their life cycles. Throughout the southwest many riparian areas were dominated by large cottonwood trees. The loss of spring flooding due to extensive damming and agricultural use has led to the spread of non-native saltcedar trees (*Tamarisk*), further preventing the native cottonwood's ability to replace themselves. Without the return of natural flow variability to our rivers, riparian areas, including the famous Bosque along the Rio Grande River, is like the living dead. The large cottonwood trees will slowly die over the next few decades only to be replaced by saltcedars and Russian olives (*Elaeagnus angustifolia*). In this case, the loss of natural disturbance regimes is leading to the spread of invasive species, which is causing a succession in the dominant plants and animals that live along the rivers and their riparian areas.

The spread of invasive species can also cause drastic changes to ecosystems. Once established invasive species displace the native species, causing shifts in the communities, often accompanied with further losses of diversity. One hundred years ago, the spread of a pathogenic fungus called *Cryphonectria parasitica* caused the chestnut blight that wiped out the American chestnut tree (*Castanea dentata*), which was the dominant tree of the entire Eastern forest. The loss caused a major restructuring in the canopy and directly caused the extinction of several animals. In southern Florida *Melaleuca* trees (*Melaleuca quinuenervia*) were intentionally spread throughout the everglades in the 1920s. Over the past 90 years they have grown into large trees forming very species poor-forest where once thriving wetland marshes existed. In aquatic communities, the rise of nutrients has helped to spread *Hydrilla* and other invasive plants into our lakes and rivers. Once established, *Hydrilla* displace the native vegetation and alters fish communities, most of the time skewing them to much smaller fish, much the chagrin of local fishermen.

Similar Climates and Ecosystems
Form a Biome

Worldwide, there are different biomes based on similarities in climate and ecosystem processes. Biomes are large regions that have common characteristics such as rates of primary productivity and patterns of diversity. There are both aquatic and terrestrial biomes found throughout the world. Although similar biomes on different continents may share very few species in common, those species must be adapted to survive in similar environments. Take a tropical rainforest for example, it has hot, humid conditions and intense competition between species. Many types of temperate trees found in North America could probably live in a tropical rainforest (they have a large fundamental niche), but they are not found there because they would be outcompeted by fast growing species that don't have to shed their leaves for part of the year (they have a smaller realized niche due to competition).

In this section, I will cover seven terrestrial biomes; tropical rain forests, temperate forests, coniferous forests, savannahs, grasslands, tundra, and deserts. As a note, other ecologists may recognize different types of biomes, such as chaparral.

Tropical Rainforests

Tropical rain forests are located near the equator and have the highest rates of annual transpiration (water coming from plants), the highest rates of primary productivity, and are among the most diverse ecosystems on the planet. The tropics are exposed to direct sunlight, every day for twelve hours, all year long. In these regions, hot air rises and it cools causing the water to condense and fall back as rain. Therefore, tropical rain forests have a continuous growing season all year long and are not resource limited. Coinciding the with the large energy inputs, the rates of evolution and speciation are also quite high.

Primary productivity in rain forest is over twice the rate as in temperate forests of the eastern U.S. The diversity in the tropics is staggering. Take the small country of Costa Rica, at 19,700 square miles it is less than the size of West Virginia and only 1/6 the size of New Mexico. Yet, it has over 800 species of birds, more than twice what is found in New Mexico and roughly equal to the number of birds found in all of the United States with an area of 9.6 million square miles. Tree diversity is also quite high, in regions of the Amazon rain forest in Brazil, over 300 species of trees have been identified in one acre (approximately 43,000 square feet). Once again, in all of North America, there are about 1,000 species of trees. To say that estimating insect diversity in the tropics is difficult, would be an understatement. There could be upwards of a million of unknown insect species in the tropics alone.

Recent studies in Ecuador have found over 100,000 species per acre solely in the canopy of a virgin forests. To put that in perspective, the Great Smoky Mountains National Park in North Carolina and located in a temperate biome has an estimated 50,000-80,000 species in its 522,000 acres or 816 square miles.

Tropical rainforests also have some of the highest rates of evolution on the planet. Intense competition among species combined with a year long growing season, and lots of energy have been cited as reasons for the rapid rates of evolution. Some research has also indicated that numerous mutual symbiotic relationships, or the cooperation between organisms could also be a major factor driving the rapid rates of evolution helping to explain the incredible species diversity of the tropics.

Not everyone totally agrees in how to define a biome. Along the northwest Pacific coast of North America is another type of rainforests, dominated by large conifer trees. These regions being farther north receive much less energy, but plenty of rainfall. They are not as diverse as the tropical rainforest, but are home to Redwoods and Sequoias, some of the largest trees on the Earth. Some scientists classify this rainforest in northern latitudes as temperate rainforests to separate them from the tropical rainforests.

Temperate Forests

Temperate forests are found at mid-latitudes, mostly in the northern hemisphere. In the summer, they strongly resemble tropical rainforest in being hot and humid with high rates of primary productivity. These ecosystems are mostly dominated by large broadleaf deciduous trees (trees that lose their leaves on a seasonal basis) and temperate conifers including pines and hemlock. During the summer, they are home to numerous song birds and insects are abundant. By late fall, the temperatures begin to drop and the deciduous trees begin to lose their leaves greatly reducing primary productivity. Many insects lay their eggs and die at the end of the summer season, while others find places to safely overwinter. Nearly 5 billion songbirds will migrate southward to the tropics to avoid the winter. Temperate forests can be quite diverse in the summer, but they all undergo the seasonal reductions during the cold winters.

Northern Coniferous Forests

The northern coniferous forests forms the largest continuous forests in the world as it stretches across Europe, Asia, and North America in Alaska and Canada. During the last ice age, the forests in North America were connected to Asia and today they are only separated by the narrow Bering Sea. These forests are dominated by cold hearty evergreen conifers that keep their needles all winter long. The trees are often narrow with short branches to prevent snow from accumulating and breaking them in the winter. They also have other adaptations to prevent the needles from freezing in the cold winters.

Summer is short lived, lasting only a few months when temperatures are relatively warm and the region comes alive with numerous insects, songbirds, and active mammals mating and storing fat for the long winters. The northern coniferous forests may have a large standing crop despite their low primary productivity. Standing crop is defined as the amount of biomass in a certain area. In coniferous forests the trees may be less than a hundred feet tall, but could be hundreds of years old, adding only a little biomass each year during their short growing season.

Tropical Savannahs and Temperate Grasslands

Savanna and temperate grasslands are dominated by grasses as their name suggests. Trees and other woody vegetation are sparse in these regions for both biotic and abiotic reasons. These biomes typically have a growing and a non-growing season, although the cause of the two seasons is for different reasons. Savannahs are found in tropical regions where seasonal variability is determined by rainfall. The dry season limits productivity and frequent fires combined with herbivory remove biomass. Temperate grasslands have cold winters with reduced productivity. Because savannahs remain warm all year long, compared to temperate grasslands, some ecologists consider them to be different biomes. Both of these biomes rely on grazing and fire to maintain their diversity and prevent encroachment of woody vegetation.

Temperate grasslands and tropical savannahs lack a large standing crop even though primary productivity can be quite high at times. To compare, coniferous forests to the north have a much higher standing crop, but their primary productivity is much lower. It turns out that disturbance is an important component of grasslands and savannahs. Routine fire every few years combined with intense grazing from mammals removes large amounts of biomass and helps to recycle to the nutrients. Many grassland ecosystems would actually become much woodier if not for the fire and grazing pressure preventing the encroachment of trees and other woody vegetation.

Tundra

To the far north or above the tree line of tall mountains is the tundra, a biome dominated by low growing plants, lichens, and mosses. The growing season is very short, restricted to only a few months in the summer when the snow melts. In the tundra, trees are greatly limited in size or are non-existent, their presence is restricted by the brutal cold of winter months.

Arctic tundra is located north of the tree line and much of it lies above the Arctic Circle, where the sun shines continuously for 24 hours every day during the summer. Shorebirds routinely migrate to breed in the Arctic tundra, only to leave once the temperatures begin to drop as the days become shorter. In the summer, only the first few inches of the ground thaw, the rest remains permanently frozen and is called permafrost, which is a major factor limiting the spread of trees northward. Over thousands of years, plant material has accumulated as frozen peat, in some cases, the peat can be thousands of feet thick. Because it is frozen, bacteria and other decomposers are not able to break down the organic material, so it slowly accumulates forming large peat deposits.

Many mountains are high enough reach above the tree line where tundra can be found. In the Rocky Mountains, it's called alpine tundra to differentiate it between the Arctic tundra to the north. Once again, the tree line is mostly determined by cold temperatures. As you go up a mountain, the temperature drops about 3.5 F for each 1,000 feet in elevation gain. In New Mexico, the tree line is about 11,500 feet, which means the average temperature at this elevation is 38 degrees colder than if you were at sea level. However, the tree line becomes progressively lower as you go northward. Alpine tundra typically lacks the large accumulations of peat found in the Arctic Tundra, but still has numerous wildflowers that quickly bloom and seed before the cold winters.

Deserts

Deserts are known for being dry, some receiving less than 10-inches of rain per year. They cover nearly 20% of land's surface and are found on every continent, with the majority found around 30-degrees latitude. Large atmospheric circulation patterns are mostly responsible for the distribution of deserts. However, local conditions, such as the presence of mountain ranges are also important for creating deserts.

We are most familiar with hot deserts, including the Sahara Desert of Northern Africa, or the Sonora Desert in the U.S. Plants and animals living in these deserts must be adapted to the hot dry environments, especially during the summer months. The harsh conditions created by high temperatures and a lack of precipitation and moisture make living difficult for most organisms to survive. Water availability totally limits plant productivity. As a result, there are few large mammals and vegetation can be scarce to almost totally absent. Deserts can be cold, such as Antarctica or the Great Basin Desert of Nevada and Utah. Deserts in Antarctica are extremely dry due to cold air that holds less moisture and mountains blocking the movement of moisture into certain areas of the continent.

The southwestern U.S. is dominated by four deserts, the Sonora, Chihuahua, Mojave, and the Great Basin Desert, which is considered a cold desert. In New Mexico, the Chihuahua Desert occurs in the southern third of the state. It is a relatively mild hot desert due to its higher elevation. It receives about 8-16 inches of rain per year, mostly in July through September from Monsoonal rains. The dominant plant is the nearly ubiquitous creosote bush, but there are also numerous species of cacti. Arizona is home to the Sonoran Desert famous for its large columnar cacti. Parts of western Arizona and eastern California form the Mojave Desert famous for its Joshua Trees.

Marine Ecosystems

Marine ecosystems are the most common ecosystems on the Earth. Approximately 70% of our planet is covered by oceans with an average depth of nearly 2.4 miles. Light only penetrates to about 600-700 feet in the open ocean. Therefore, the most common habitat on the planet is the aphotic zone and the abyssal plain. Many of the animals living here, never see sunlight in their life. These deep sea organisms can appear quite alien to us. Some are capable of producing their own light through bioluminescence to attract prey which they catch with their enormous mouths full of long teeth. These deep see communties actually have very few if any autotrophs. Instead, they receive the bulk of their energy as planktonic snow, that is the slow but steady trickle of dead organisms that float down from the surface to the depths of the ocean. Some deep sea organisms make nightly migrations to within a few hundred feet of the surface to take advantage of more resources. In recent years, it has been discovered that whale carcasses form important habitats and sources of energy for the abyssal plain.

Thermal vents are found in the depths of the oceans where the continents are being pushed apart by powerful geological forces. The ecosystems surrounding the thermal vents are teaming with life, yet not a single organism depends on energy from sunlight. These unique communities were discovered in the 1970s and represented an entirely unknown and unique ecosystem dependent solely on chemosynthesis. Here, small autotrophic bacteria extract energy directly from the minerals spewing out of the vents and serve as the base of the food web. Hot vent communities may only last a few decades before they become inactive and the inhabitants must migrate to new active sites.

Perhaps the exact opposite of a thermal vent are the coral reefs. Found in warm shallow waters, these are among the most diverse ecosystems in the world. Although, they occupy less than one percent of the ocean's surface, they may account for nearly half of its diversity. Coral reefs are dependent on the corals that build structure by excreting calcium carbonate from the ocean water.

Kelp forest are very productive ecosystems found mostly near the coast from temperate to polar climates around the world. Kelp is actually a general term for several species of large algae where some can grow over 100 feet in length creating a canopy. These large canopies provide important habitat for many other species of fish and invertebrates.

Also found in shallow waters close to shore are seagrass beds. These ecosystems are dominated by flowering plants that superficially resemble grasses because they have long narrow leaves. Many fish living on coral reefs will visit seagrass beds to feed at night. Seagrass beds have very high rates of primary production and serve as nursery grounds for many species, inlcuding several popular sport fish and bay scallops.

Chapter 13

Climate Change

Science is not political,
if I tell you that more carbon dioxide in the atmosphere
will raise the Earth's temperature,
it is borne out of observations and experimentation
that can be repeated.
It is not an ideological assertion based on a belief system.

Introduction

Life is intricately connected to geological and atmospheric processes. For over 3.5 billion years, life has been slowly altering the composition of the atmosphere. When life first emerged, the atmosphere contained very little oxygen, perhaps less than a thousandth of a percent. Instead, it was likely dominated by carbon dioxide and nitrogen. The air we breathe today is about 22% oxygen, a product of photosynthesis, a process of life. In fact, if scientists were to discover another planet with an oxygen rich atmosphere, it would be a smoking gun for life.

As the composition of gasses in the atmosphere has changed over time, so has the climate. Before I paint a picture that climate and the atmosphere are solely influenced by life, climate is also influenced by other physical factors as well, including volcanism and the position of the continents. But one thing is certain, when the climate changes, it has an enormous impact on life. Species go extinct, new species evolve, entire ecosystems may collapse while others emerge.

The Earth's climate has a history of change as it oscillated between hothouse and icehouse climates. Change has occurred slowly over the course of millions of years at the pace of continents moving across the globe, to periods of rapid glaciation or melting that take place over centuries. Rapid climate change has been implicated in several mass extinction events.

Once again the Earth's climate is beginning to rapidly change. Although, this time it's different, it is caused by our activities, mainly the burning of fossil fuels that is releasing billions of tons of carbon dioxide into the atmosphere every year, an unpresented rate not seen in the Earth's past. Numerous lines of evidence from the loss of glaciers and sea ice, rising sea levels and record temperatures, clearly indicate we are warming the planet. By refuting natural causes of climate change, we know with a high degree of certainty that it is our activities causing global climate change, which will have a huge impact to all life and our civilization.

An Introduction to Climate

The Earth's climate is complicated, there are numerous factors that keep it stable, or are capable of causing it to change, and at times drastically. Before diving into climate, let's begin by clarifying the difference between weather and climate. Weather can be described as the daily activity of the atmosphere. Weather is what is happening right now; is it rainy, snowy, sunny, or windy, hot, or cold. And weather can change rapidly, an oncoming thunderstorm can drop the temperature by 10-20°F in less than an hour. Local weather events are influenced by many variables making it difficult to forecast beyond few days with any accuracy.

Climate, on the other hand, is the average weather you would expect for a region at a specific time. When you watch a weather forecast and the meteorologist claims the high temperature for the day is above average, they are comparing that day to a 30-year average for that day. The climate of any one region is determined by a complex mix of local geology, such as the presence of mountain ranges, the position of the continents, the latitude of a region, large scale global circulation patterns, history, and of course the composition of gasses in our atmosphere. Tropical rainforests experience a climate with constant warmth, and average daily highs and lows may fluctuate less than a few degrees throughout the year. Temperate forests may have a 40°F difference In average seasonal temperatures.

The atmosphere blankets the Earth and together with ocean currents they work to redistribute energy around the planet. If it weren't for the atmosphere, daytime temperatures in the tropics would soar hundreds of degrees during the day only to plummet to -200°F at night. The poles would remain frigid with temperatures remaining less than -200°F. Almost certainly, life would have a very difficult time existing without the atmosphere to redistribute the incoming solar radiation and mitigate day and night temperature differences by acting as a natural greenhouse.

The composition of gases in the atmosphere is a major factor influencing how much heat is retained. Currently the composition of our atmosphere is approximately 78% nitrogen, 21% oxygen, and about 1% trace gases including carbon dioxide, methane along with some water vapor. Water vapor, carbon dioxide, and methane are the most abundant greenhouse gases because they retain heat in the atmosphere. You may have noticed how night time temperatures remain higher on cloudy nights, or when the air is humid. The water vapor in the atmosphere retains heat from the day and prevents it from escaping into space at night. Carbon dioxide and methane also trap incoming solar radiation. Even though they are trace gases, small changes can lead to large changes in the climate. We are currently witnessing this phenomenon right now as increased carbon dioxide in the atmosphere is causing the current rise in global temperatures.

A History of Climate Change

Throughout Earth's long history, the climate has fluctuated between episodes of extreme ice ages with extensive glaciation to periods of warmth known as a 'hothouse' where there were no continental glaciers or polar ice caps. Sometimes, climate change would remain stable for millions of years, whereas other times the climate would change relatively rapidly.

Reconstructing the past climate becomes more challenging the further back in time you go. Geological processes constantly erode the surface as rocks are recycled, forever removing pieces of evidence for Earth's deep past. Today, only a few places have rocks that are over 3.5 billion years old. There are basically three types of evidence allowing us to reconstruct ancient climates for these ancient ice ages including geological, chemical, and paleontological. Closer to modern times, we have additional sources of information including ice cores, sediment cores, and tree ring data available to accurately reconstruct the past climate.

For much of the Earth's history, atmospheric processes, geological processes, and life's processes have been intricately connected, they do not exist in isolation of each other. The Earth's climate is strongly influenced by the composition of gases in the atmosphere, which in turn, is a product of geological and living processes. Prior to the origins of life, the atmosphere was influenced by geological processes and to some extent, large meteor strikes and solar output. Based on our best evidence, the atmosphere 4 billion years ago was an alien atmosphere, a potential mix of methane, carbon dioxide, nitrogen, and water. It was unable to support animal life because oxygen was almost completely absent. Carbon dioxide and methane, were likely much more prevalent, which was a good thing because the sun was about 25% cooler than it is today. It is thought that the high levels of these two greenhouse gases, along with water vapor, kept the Earth warm and prevented the oceans from freezing.

Recent theories have suggested that both living organisms and geological processes may have stabilized the atmosphere early on, preventing runaway greenhouses like what happened to Venus, or preventing the loss of our oceans like what happened to Mars. I should also point out that Mars is also dry due to its smaller size and lack of a magnetic field to protect it from the solar wind. Although the Earth's climate has not gone to quite the extremes in climate as our two nearest planetary neighbors, the Earth's climate has undergone major swings when global temperatures may have fluctuated by nearly 27°F (15°C). At the cold end during severe ice ages, global average temperatures were about 50°F (10°C) and during warmer periods, global average temperatures may have spiked above 77°F (25°C). Although, 77°F may seem warm, temperatures in the tropics could have easily reached 120°F and some data indicate ocean temperatures on the surface may have reached about 100°F. In comparison, the current average global temperature is about 58.7°F (14.8°C), and surface ocean temperatures in the tropics are around 80-83°F. Each time the climate changes, it has had a profound impact on the diversity of life and its future evolution. Based on different types of evidence there have been at least five major ice ages lasting from a few million years to hundreds of millions of years.

The Proterozoic Ice Ages

Evidence for the first major change in climate due to life dates back to at least 2.4 billion years ago to the Proterozoic Eon. It's thought that the Great Oxygen Event may have triggered the first major ice age. Ancient cyanobacteria continually pumped oxygen into the atmosphere for over a billion years, slowly altering the composition of gases in the atmosphere. The working theory is that the oxygen produced from photosynthesis removed methane from the atmosphere. Over time, the loss of methane cooled the Earth and it entered an ice age lasting nearly 300 million years. It is thought that continual volcanism released carbon dioxide into the atmosphere, eventually warming the climate and ending the Earth's longest ice age.

Once again about 720-635 million years ago, the Earth entered into another ice age called the Cryogenian Period. This ice age may have been so severe that extensive glaciers on land reached the tropics and sea ice extended almost to the equator covering much of the planet. The extent of the glaciation may have been so drastic that it has been called the "Snowball Earth". Some scientists question the extent of the glaciation and call it the "Slushball Earth". So, how did the sequestration of carbon into geological cycles far outpaced additions to the atmosphere?

The Earth has always had active volcanoes serving as a natural source of carbon dioxide into the atmosphere. Luckily, carbon dioxide has never built up to catastrophic levels as seen on Venus where surface temperatures are hot enough to melt lead. Part of the reason why that has not happened on Earth is that carbon dioxide is constantly removed from the atmosphere by the chemical weathering of rocks and the action of photosynthesis. About 700 million years ago the continents drifted near the equator forming a super continent called Rodinia. Being subject to higher rainfall, the continental rocks would have undergone higher rates of chemical weathering, thus removing carbon dioxide from the atmosphere as it became sequestered into carbonate rocks. The removal of carbon dioxide from the atmosphere would have lowered the Earth's average temperature causing the spread of glaciers.

Through positive feedback the glaciers would have continually grown, eventually reaching the tropics. It may have gone something like this: Less carbon dioxide would have caused colder winters leading to more ice cover. Then, cooler summers would have resulted in less ice being melted, leading to more ice being formed in the winter. More ice would reflect additional energy back to space, thus further cooling the planet, once again leading to more ice in a positive feedback cycle leading to a runaway ice age for nearly 100 million years. But, it was not to last forever.

Slowly, carbon dioxide levels would rise enough to warm the planet. Photosynthetic organisms are a major force removing carbon dioxide. About 700 million years ago when the Earth was in a runaway ice age with ice covering much of the oceans, the removal of carbon dioxide by photosynthesis was greatly reduced. The glaciers also covered the continents, preventing the chemical weathering of rocks, further slowing the removal of carbon dioxide. During this time, volcanism never stopped, so carbon dioxide levels in the atmosphere slowly rose over millions of years. Eventually, carbon dioxide in the atmosphere began to accumulate to levels that would trap enough heat to break the grip of ice age.

Eventually the climate warmed and about 600 million years ago the glaciers retreated. In yet another example of a positive feedback the ice age ended; as the glaciers retreated, less energy was reflected to space, rather it was absorbed by the oceans causing the glaciers to further retreat. Eventually, Rodinia began to break apart when the continents drifted southward, which reduced chemical weathering allowing carbon dioxide to continue its slow build up in the atmosphere. By the end of the Proterozoic Eon 542 million years ago, the Earth was quite warm and carbon dioxide levels were 5 times higher than today.

The Phanerozoic: Hothouse - Icehouse

It's been hypothesized that the intense ice age ending 600 million years ago may have drastically slowed the evolution of life, delaying the appearance of the first animals. Indeed, the start of the Cambrian Period, which is defined by the appearance of abundant animal fossils (specifically the first appearance of trilobites), took place nearly 55 million years after the end of the intense glaciation. For much the last 542 million years of the Phanerozoic Eon, the Earth's climate has gone from warm and wet periods to ice ages much like the one we are in today. But nothing as intense, or as long as the global glaciation during the Proterozoic Eon.

The Phanerozoic Eon remained warm for nearly 100 million years until its first ice age gripped the planet nearly 450 million years ago. It's been thought that during this time the continents were in the southern hemisphere and moved over the South Pole causing massive glaciation. The average global temperatures dropped about $18^{o}F$ ($10^{o}C$) to an average temperature aroung $55^{o}F$ and caused a mass extinction of early sea life. Luckily, this was a short-lived period of glaciation and the glaciers were much less extensive than the previous Snowball Earth. Over time, the continents slowly drifted away from the South Pole and surface temperatures once again returned to their previous levels.

The Devonian Period 400 million years ago was a unique time in the Earth's history. The climate was very warm and the land masses were once again near the equator. In the oceans, fishes were becoming quite diverse, some even refer the Devonian Period as the *Age of Fishes*. On land, plants were evolving into the first forests comprised of giant fern trees forming vast swampy forests. Arthropods had colonized the land and primitive insects were evolving flight for the first time. By 300 million years ago, oxygen levels were much higher than today, accounting for 30% of the atmosphere, allowing arthropods to grow much larger than today. Recall that the oxygen in the atmosphere comes from photosynthesis and is removed by rock weathering. During the Carboniferous, the Earth was covered in vast forests pumping lots of oxygen into the atmosphere. These forests covered the surface, preventing rock weathering, as a result, both oxygen and carbon dioxide levels soared.

The spread of ancient forests may have initially caused carbon dioxide levels to rise, but they also caused them to drop later on. Over millions of years, the ancient forests slowly removed carbon dioxide from the atmosphere. As the plants died and fell into the swamps, bacteria were unable to fully break down the organic matter resulting in large deposits of coal and oil. Overtime, carbon dioxide levels fell from 2,000 ppm to about 250 ppm and the world once again entered into an ice age. It was followed by a minor extinction event, although not one of the Big five mass extinction events. The slow change in climate spurred the evolution of seed plants in response to the cooler, drier climates.

About 251 million years ago, near the end of the Paleozoic Era, carbon dioxide and temperatures rapidly increased due to extensive volcanism in present day Siberia. The intense volcanism not only changed the atmosphere, it made the oceans much more acidic, causing the largest mass extinction of the Phanerozoic Eon. It has been estimated that 80-95% of all life perished in less than a million years, a very short time in the geological record.

The End Permian extinction marked the end of the almost 300 million years of the Paleozoic Era and ushered in the Mesozoic Era. The start of the Mesozoic Era may have been very hot, perhaps the hottest the Earth had ever been with some evidence suggesting that surface ocean temperatures may have reached 104°F. After about 20 million years, the average global temperatures cooled, but still remained high for much of the Mesozoic Era, known for the reign of the dinosaurs. However, beginning 100 million years ago carbon dioxide levels began to slowly decline. By sometime around 35 million years ago, the Earth's climate began to cool and dry once again. Eventually polar ice caps began to form about 34 million years ago. By the start of the Pleistocene Epoch 2.6 million years ago, the Earth entered into another ice age that we are still in today. Although some scientists think that we have been in an ice age for 34 million years, placing the start with the freezing of Antarctica. Antarctica has not always been covered with large glaciers, prior to 35 million years ago, it was warmer and forested.

For the last 450,000 years, and perhaps 2 million years, the Earth has oscillated between periods of extensive glaciation and periods of reduced glaciation known as interglacials. Currently, we have been in an interglacial for the last 10,000 years. Several of the interglacial periods have been warmer by approximately 3.9°F (2°C). This temperature difference may not seem like much, but it corresponds with sea levels that were about 20 feet higher than present sea levels. When you look at the cyclic nature of climate change over the past 450,000 years, the tight relationship between carbon dioxide levels, global temperatures, and sea level becomes apparent. For the last 10,000 years, we have enjoyed a period of relatively stability in the climate that has seen the unprecedented rise in human civilization.

Evidence for Current Climate Change

Science is not political. The process of science is dependent on making observations, formulating hypotheses, testing hypotheses, and keeping ones that fit the data while discarding hypotheses that do not fit the data. Science is an iterative process, which builds on previous knowledge. Over time our knowledge changes so that our world view is a reflection of our current state of knowledge. The same holds true with our knowledge of climate change. Let's explore the evidence of modern climate change based on scientific observations. I use the term scientific observations to make it clear that the data is accurate and can be observed or replicated by anyone who's looking.

There are numerous lines of evidence supporting modern climate change going back to Arrhenius in the 1890s. However, I'm going to begin in the late 1990s when Michael Mann and others reconstructed the climate going back to approximately 1000 years. The goal was to accurately reconstruct the past climate to determine the causes of climate variability. To reconstruct the past climate, they used proxy data from tree-rings, corals, sediment cores, and ice cores along with other forms of "proxy" data going back about 1000 years. For the past century, they included actual temperature measurements to the model, which were quite revealing of the rapid rise in temperatures in the 20th century.

With the current temperature data included in the model, it became

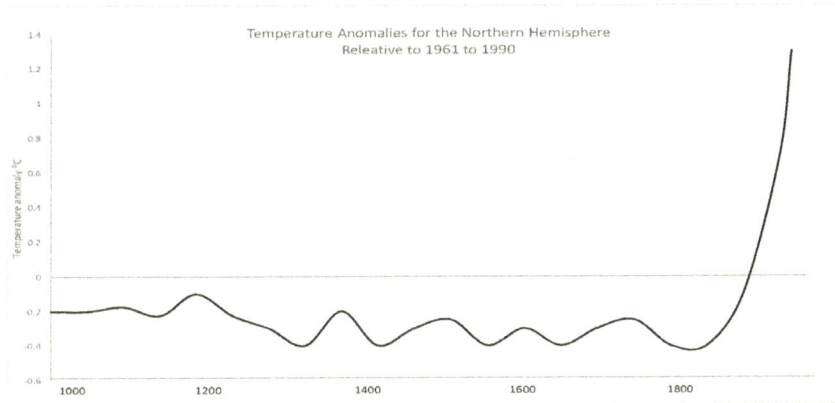

Temperature Anomalies for the Northern Hemisphere
Releative to 1961 to 1990

A graphical representation of the Hockey Stick Model of climate dating back 1,000 years based on proxy data Generated by Mann *et al.* 1998. For approximately 900 years, the climate was relatively stable, but in the last 100 years, average temperatures have risen about 1.3 degrees Celcius. This graph includes the latest temperatures from 2016, to further emphasize the rapid rise in global surface temperatures.

commonly known as the Hockey Stick Model of climate change due to its resemblance of a hockey stick. There are several other conclusions easily drawn from the graph. First, the past climate was relatively stable. There were warmer and cooler periods, although the fluctuations were less than 1°F (0.5°C). Second, the average temperatures for the last 1000 years were below the average temperature set between 1961 and 1990. Third, the average temperature increased nearly 1.25°F (0.7°C) from 1900 to 1998. coinciding with the industrial revolution and the burning of fossil fuels. They also showed that the 1990s were the warmest decade in the last thousand years, and 1998 was the warmest year of the millennium. And lastly, the rapid rise in temperature of the last 100 years corresponded with the rise in carbon dioxide in the atmosphere from the burning of fossil fuels.

Surprisingly, the validity and accuracy of the peer-reviewed graph was questioned by non-scientists, politicians, and industry specialist. To be blunt, the anti-climate change segment of society attempted to discredit Dr. Mann and his work. However, scientific results are available to everyone so the methods can be repeated and the results can be independently verified. After the political attacks to the Hockey Stick Model, other scientists set out to verify it by duplicating his work. Within a few years, other scientists repeated and expanded on Dr. Mann's work only to find the same climate pattern of rapid temperature increases in the last century. Their findings exonerated Dr. Mann and his work, further lending support that Earth's average temperature is rapidly rising. In fact, the rapid rise of temperatures of the last 100 years is unparalleled in the geological record. It may be the fastest rate of temperature increase ever observed.

Since the publication of the Hockey Stick Model, we have had the warmest decade on record from 2001-2010, surpassing the 1990s. The record heat of 1998 has been broken several times, 2014 was the warmest year on record only to be broken again by 2015. The record heat of 2015 also coincided with a super El Nino which broke the one-week record for warmest sea surface temperatures. July of 2016 was the warmest month ever recorded and January 2016 broke the record for the largest temperature departure for a month. According to NASA, as of July 2016, the average global temperature was 2.4°F (1.3°C) above the late 1800s. In less than 20 years, the world almost doubled the amount of warming witnessed from 1900-1999.

In addition to the rise in surface temperatures, scientists have observed that the Earth's cryosphere, the frozen part of the planet, has been rapidly declining for the past 50 years. Starting in 1979, NASA sent up satellites to continuously monitor the seasonal variations in Arctic sea ice, providing a much more complete picture than previous records made based on location.

In 1979, NASA launched a satellite to track the Arctic sea ice. Since these first direct satellite observations, the Arctic sea ice has been rapidly declining at about 13.4% per decade. While 2016 did not set a new record low for Arctic sea ice, it tied 2007 for the second lowest extent square miles below the 1981-2010 average, or about 40% less area than the late 1970s. In addition to the loss of Arctic sea ice, the Greenland ice sheet has also experienced warmer than average temperatures leading to increased melting of the glaciers. In fact, worldwide, many glaciers have been rapidly retreating. Glacier National Park is a case in point. In 1850, the area that forms the park had 150 active glaciers, today it only has 25 active glaciers. In a stable environment, you would expect some glaciers to be retreating, some to be stable, and others growing. Instead, we see about 85-90% of all glaciers are retreating worldwide.

In the 1990s, computer models created by climatologists began to predict that climate change would cause an increase in severe weather. This intensification of weather patterns would mean intensification of persistent drought, flooding, more intense storms, and an increase in more hurricane and cyclone activity. Although, any one particular storm may not be directly caused by climate change; however, an increase in extreme weather events can be attributed to climate change. For example, California has been caught in a long-term persistent drought starting in 2012 and has continued through 2016 despite an El Nino. There was another intense drought across the Midwest in 2012 greatly reducing crop production. In 2015, the eastern Pacific in the El Nino region recorded its highest temperature ever, surpassing the El Nino of 1997.

Taken together, the rise in surface temperature, loss of the cryosphere, and the increase in extreme weather events all point to a rapid change in the Earth's climate. In fact, the current rate of climate change is among the fastest ever known to occur. There is no information supporting the idea that the climate is not changing, all the data from numerous observations around the world clearly indicate the climate is changing. Based on these observations, climatologists have focused their attention on studying the factors that caused previous climate change and how the Earth will be affected by current climate change.

The Natural Causes of Climate Change

For more than a century scientists have been reconstructing the past climate to determine the causes of climate change. It's important to do so because by understanding the past, we can understand the present.

As you just learned in the previous section, the Earth's climate is a very complicated system where even small changes can lead to new climate regimes. Natural causes of climate change are also quite varied, they can include the movement of continents, periods of intense volcanism, changes in the Earth's orbit, changes in solar output, meteor impacts, and the processes of life. And each of these factors can amplify or mitigate each other. But, the take-home message is that many of these factors affect the composition of gases in the atmosphere, which leads to global climate change.

Using several different types of data, scientists have accurately reconstructed the past climate going back approximately 850,000 years. The results clearly show we are currently in an ice age and our climate has oscillated between periods of maximum and reduced glaciation. Over this time, the climate fluctuated about 9°F (5°C) and carbon dioxide levels fluctuated between a low of 185 ppm to around 300 ppm over periods of hundreds of thousands of years. Another pattern noticed by scientists is that warming was quite rapid and lasted for shorter time periods than the ensuing periods of maximum glaciation. What were the natural causes driving the cyclic nature of our climate for the past 850,000 years and how could they be affecting us now? For the remaining part of this section, I cover the major natural causes of climate change and show how scientists can rule them out as causes for the current climate change.

Milankovitch Cycles

One explanation for the climates' cyclical pattern are slow changes in the shape of the Earth's orbit on scales lasting 26,000 to 100,000 years, known as Milankovitch cycles. There are three different cycles, eccentricity, obliquity, and precession. The shape of the Earth's orbit is known as its eccentricity and it changes over time from nearly circular to more elliptical, each cycle lasts about 100,000 years. The tilt of the Earth's axis known as obliquity varies between 22.5° and 24.5° over a period of 41,000 years. The tilt of the Earth's axis is currently about 23.5° and is

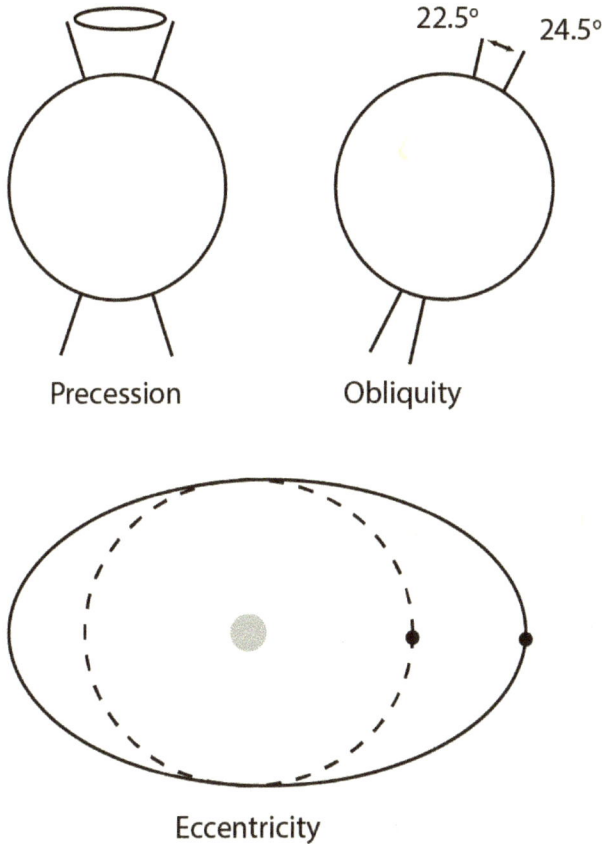

22.5°   24.5°

Precession   Obliquity

Eccentricity

Milankovitch cycles include the eccentricity of the Earth's orbit, the obliquity, which is the tilt of the axis, and precession which is the wobble on of the axis.

getting smaller. Precession is the wobble of the Earth's axis, or changes in the direction of the axis. Today, the North pole points at the North star called Polaris, but thirteen thousand years ago, the North pole pointed at another star. It takes about 26,000 years for this cycle to complete itself.

The interplay of these three cycles can affect the global climate. Based on the current Milankovitch cycles, tilt of the axis is actually decreasing and combined with a nearly circular orbit, the Earth should actually be slowly cooling. Therefore, scientists have ruled out Milankovitch cycles as a cause of any current warming trends.

Volcanism

Periods of intense volcanism can temporarily cool the planet by emitting large quantities of ash and aerosols into the atmosphere. In 1991, Mt Pinatubo exploded in the Philippines releasing 20 million tons of sulfur dioxide into the atmosphere and temporarily reduced global temperatures by approximately $1^{\circ}F$. However, long-term increases in volcanism can release large amounts of carbon dioxide, raising atmospheric concentrations enough to warm the planet. At least two of the last three mass extinctions have been blamed on periods of intensive volcanism, including the End-Permian 251 million years ago and potentially another round of volcanism 201 million years ago, causing the end of the Triassic. There is also evidence of yet another major round of volcanism in modern day India known as the Deccan Traps, which may have been a factor leading to the extinction of the dinosaurs. Today, there has not been any major increase in volcanism or geological activity that would lead to greater amounts of carbon dioxide in the atmosphere.

Meteor Impacts

There are factors outside the Earth that can alter the climate. Large meteor impacts can drastically alter the climate in a day. It is believed that a large meteor hit the Earth 65.5 million years ago and caused such a rapid change in climate that it wiped out the dinosaurs. However, there is some support that the meteor impact was preceded by a period of intense volcanism in present day India creating the Deccan Traps that may have caused severe declines in dinosaurs prior to the meteor impact. There have been no recent meteor impacts significant enough to cause any current climate change.

Solar Output

Over the last 4.6 billion years, solar output has changed drastically. Early in the Earth's history, the sun was less intense, emitting only 70% of the energy it does today. In the far future, over hundreds of millions of years, the sun will continue to increase in brightness as it burns through its hydrogen.

On shorter timescales, the sun undergoes an 11-year solar cycle when solar output cycles from periods of low to higher intensity. There may be other longer-term cycles including a 90-year cycle, but they are not as well known. One of these longer cycles may be responsible for the Maunder minimum, which are periods of reduced sunspot activity that correspond to lower solar output. The lower solar output with fewer sunspots corresponded to the Little Ice Age when European and North American temperatures were below average between 1645 and 1715. Current observations of the sun have shown that we have been in a period of lower solar output that may intensify in the next few decades. Based on solar output, the climate should be cooling. Therefore, we can rule solar activity for the current warming we have been observing.

Position of the Continents

The position of the continents can help create the conditions for an ice age. One way is when the block the movement of warm water currents to the poles. For example, Antarctica sits on top of the south pole and is the coldest place on the planet. The Arctic is mostly landlocked by North America, Europe and Asia that prevent the movement of warm water currents into the region. Rodinia may have caused the Snowball Earth 700 million years ago by sitting at the equator, which may have prevented warm water currents moving to the poles. Also, being at the equator, it would have had high rates of rock weathering, acting as a giant carbon dioxide scrubber, reducing carbon dioxide levels.

The current ice age may have been triggered, in part, by the movement of the continents preventing warm water moving to the poles and allowing them become colder. Additional intensification of the current ice age may result of continental uplifting of the Colorado and Tibetan Plateaus that are removing large amounts of carbon dioxide from the atmosphere, in a situation similar to Rodinia. In the last 100 years, the continents have not moved sufficiently to have any warming effect on climate change. Therefore, this has been ruled out as well.

Changes in the Earth's Atmosphere

Our knowledge that carbon dioxide is a greenhouse gas dates back over 125 years ago to the French chemist Svante Arrhenius who predicted that the burning of coal could warm the climate. Greenhouse gases, such as carbon dioxide, methane, and water trap heat and prevent it from escaping back to space. The more greenhouse gases present, the warmer the Earth's climate will be. There are multiple factors that affect the amount of greenhouse gases in the atmosphere. Volcanism can add carbon dioxide to the atmosphere, whereas rock weathering and photosynthesis act to remove carbon dioxide.

In the past 125 years, atmospheric carbon dioxide levels have risen from 285 ppm to over 400 ppm, levels not seen in over 2.2 million years. There have been no major changes in volcanic activity or other natural sources that has caused the rapid increase in carbon dioxide levels.

The take home message is that there are natural causes of climate change, but they are not causing the current warming we are experiencing. Often, there may be several factors taking place at once to make the Earth hot or send it into an ice age. But most importantly, we know that when the composition of the atmosphere changes, so does the climate. Warmer climates have high levels of carbon dioxide and cooler climates associated with ice ages have low levels of carbon dioxide. And when the climate changes quickly, extinction events are likely to occur. This time, the rise in carbon dioxide is coming from the burning of fossil fuels.

Humans are the Cause of Modern Climate Change

The cause of modern climate change is actually well known; it is being driven by the rapid rise of carbon dioxide in the atmosphere from the burning of fossil fuels and to some extent, deforestation. Recall that the French chemists Arrhenius predicted in the 1890s that the burning of fossil fuels would lead to a warmer planet. Prior to the industrial revolution, carbon dioxide levels in the atmosphere were approximately 285 ppm and had been at about that level for the previous 10,000 years. The first daily monitoring of atmospheric carbon dioxide began in the 1950s at the Mauna Loa observatory in Hawaii. A climatologists named Charles Keeling wanted to test the hypothesis that carbon dioxide levels would decline in the summer and rise in the winter in response to seasonal variation in photosynthesis in the northern hemisphere.

He collected data for several years and easily detected the seasonal changes he predicted. However, in science, one discovery often leads to others. Dr. Keeling also noticed that the annual average levels of carbon dioxide were increasing. Today, the same observatory is still in operation, taking continuous measurements of carbon dioxide levels. In 2015, the average carbon dioxide levels crossed over 400 ppm, a level that has not been on Earth for over two million years or more. The last time carbon dioxide levels were that high, the oceans were 20 feet higher than they are today.

The source of the extra carbon dioxide levels is from the burning of fossil fuels. There are no observations indicating that the rapid rise in atmospheric carbon dioxide levels is from natural sources, including an increase in the rate of volcanism. There are several ways we can determine the source of the extra carbon dioxide and rule out natural sources. First, all elements come in slightly different varieties called isotopes that vary in the number of neutrons in the nucleus. Molecules of carbon dioxide from volcanic sources will have different ratios of isotopes compared to coal and oil. When scientists measure the carbon dioxide in the atmosphere, it has the tell-tale signature of coming from coal and oil. Additionally, there have been no observations of more volcanic activity compared to hundred years ago to explain the rise of carbon dioxide in the atmosphere. Also, we put about 2.2 million pounds of carbon dioxide into the atmosphere every second of every day, which translates to approximately 35 billion tons of carbon dioxide each year. Based on these numbers, it becomes clear where the extra carbon dioxide comes from.

Climatologists have also ruled out the other natural factors of climate change. For example, the continents have moved less than 10 feet since the beginning of the industrial revolution, not enough to affect ocean currents or other climatic patterns. Those kinds of changes require millions of years. Solar output does vary over time; in fact, solar output has been entering a period of low activity, which should be causing lower temperatures. Despite the lower solar activity, we have actually seen an increase in surface temperatures. And lastly, based on the Earth's orbit, we should not be getting warmer, but our climate should remain stable for a few thousand years more.

The science behind climate change follows the same methodology as any other branch of science. We make observations about the world, formulate hypotheses to explain the observations and then go and test our hypotheses. In the case of climate change, the observation is that the world is getting warmer. All of the natural sources of climate change (solar activity, continental movement, increased volcanism, and Milankovitch cycles) have been disproven by scientific observations and experiments as a cause of the current warming trends. The only valid scientific explanation remaining to explain the rise in surface temperatures is caused by our activitires. The burning of fossil fuels and clearcutting are raising atmospheric carbon dioxide levels in the atmosphere, driving the rise in surface temperatures.

Accurately Predicting Climate Change
Can be Difficult

Making accurate predictions of climate change is very complicated because there are so many variables affecting the climate. One example illustrating how it can quickly become complicated lies in positive feedback when small changes are amplified. The Snowball Earth may have been the result of positive feedback leading to global glaciation. For a modern example of a positive feedback, we can use the melting of the Arctic sea ice. Ice is mostly white, so it reflects much of the sun's energy back into space. As the ice melts, it creates more open water which is much darker, absorbing more energy, causing the water to warm. The warmer water, in turn, melts more sea ice in the summer and slows its formation in the winter. In this case, the melting of the sea ice amplifies the rate at which it melts in a positive feedback. Additionally, the loss of Artic sea ice makes the region much warmer, which in turn melts the permafrost in the tundra releasing even more carbon dioxide into the atmosphere, once again increasing the warming in the region.

Another example of the difficulties in predicting climate change lies in something as seemingly simple as clouds and cloud formation. A warmer planet will have higher rates of evaporation, holding more water in the atmosphere, potentially leading to more clouds. However, clouds are white and reflect sunlight into space, but at night they hold heat from the day. That's why cloudy days and nights have less temperature swings. However, not all clouds are created equal, it turns out that the type of cloud, its thickness, and height are very important in determining whether or not the increased cloud cover reflects more sunlight or holds more heat. Understanding the relationship between clouds and warmer climates is an active area of climate research.

Climatologists must also account for the total photosynthetic output of the planet because it removes carbon dioxide from the atmosphere and compare that to the amount of fossil fuels being burned each year. They must take into account deforestation that will add carbon dioxide and make the clear-cut areas warmer. Yet another factor to consider is that as the permafrost melts, the northern boreal forests are moving northward. As primary productivity increases, more carbon dioxide is removed from the atmosphere, but the forests are darker and absorb more sunlight than the tundra, so the region warms more. As the permafrost warms, the peat will be broken down, releasing more carbon dioxide.

Warmer temperatures also affect global circulation patterns that distribute incoming solar energy, strongly influencing regional climates around the planet. For example, a warmer Arctic in the winter can actually bring very cold air to the northern hemisphere. A polar vortex sets up as the Arctic cools in the winter, the vortex prevents much of the cool air from spilling southward. However, temperatures that are 20°F warmer in the Arctic winter weaken the polar vortex, it allows the Arctic air to spill southward. Although, the Artic air may be above average in warmth, the air is still very cold compared more southern latitudes of Europe and North America.

Climatologists look to the past to reconstruct the climate by using proxy data similar to what was used to create the Hockey Stick Model to help make predictions about current climate change and sea level rise. Taken together, these independent sources of information paint a picture of the Earth's climate through time. Geological and fossil records are useful for constructing ancient shorelines and the types of animals that were present, or went extinct. For example, when the Earth was much warmer, Antarctica was forested and ancestors to alligators lived as far north as Wyoming.

Based on the historical proxy data, we know for the last 450,000 years, there have been cycles of maximum glaciation with periods of interglacials interspersed. When the glaciers are at their maximum extent, sea levels are nearly 300 feet lower than they are today. The last time carbon dioxide levels were similar as today, the Earth was only 3.9°F (2°C) warmer and the sea levels were about 20 feet higher. These observations about sea level rise and temperature increases worry scientists about the rapid rise in carbon dioxide levels observed today. We have already warmed the Earth 2.4°F (1.3°C) over the last century, with much of that warming taking place in the last 30 years, corresponding to some of the fastest changes observed in the Earth's climate.

In the past 20 years, our knowledge of climate change has rapidly improved, along with that our ability to make predictions about its future. While there is often disagreement among scientists about which factors are most important in predicting climate change, one thing they all agree on is that climate change is rapidly underway, it is not just a future problem. There are uncertainties in our ability to predict the future, but the past patterns in the climate clearly indicate that sea levels will rise and rainfall patterns will shift. The trouble is predicting how fast the climate will change or how much sea levels will rise.

The Implications of Climate Change

What are the implications for climate change to life on the Earth? To a first approximation, rapid climate change has led to some of the most severe mass extinctions. The End-Permian Extinction 251 million years ago, brought about by a period of intense volcanism followed by rapid climate change, was likely the greatest extinction event in the Earth's history. There are numerous reasons why climate change causes extinction, especially in the oceans. When water conditions change, many forms of sea life cannot adapt rapidly enough and perish. Intense volcanism can cause rapid changes to the water chemistry. For example, much higher levels of carbon dioxide in the atmosphere causes the pH of the oceans to lower and become more acidic. The lower pH makes it increasingly difficult to extract calcium carbonate from the water to form shells or reefs, including many mollusks and corals.

Today, we are seeing the pH of the oceans lower as carbon dioxide dissolves into the water creating carbonic acid. In fact, it's been estimated that up to half of the carbon dioxide released in the last hundred years has been absorbed by the oceans. The oceans cannot continually absorb carbon dioxide from the atmosphere at the same rate, which will lead to an even more rapid rise in atmospheric carbon dioxide levels.

In the tropics, coral reefs form in crystal clear waters that rarely experience large temperature changes. The corals are able to exist in these nutrient poor conditions because they form a symbiotic relationship with photosynthetic zooxanthellae that live in the tissues of the corals. When the corals become stressed, they often lose their symbionts and take on a white appearance, hence the name coral bleaching. Higher sea surface temperatures are known to cause coral bleaching, and in severe cases, leads to death of the corals. In 2015, NOAA declared the third ever global coral bleaching event due in large part to the oceans warming and one of the largest El Nino events on record. When you consider that coral reefs are home to nearly 50% of all marine species, the implication for the loss of diversity becomes apparent.

Tropical reefs are not the only marine ecosystems being affected by climate change and ocean acidification. At the poles, warmer water not only decreases the ice cover, but it has increased the productivity of phytoplankton by an estimated 40-50% due to the longer growing season. In some ways, this is good news because it will help remove carbon dioxide from the atmosphere and could serve to boost fish populations in the region. Many marine fish can easily move into new regions as climate change results in warmer waters poleward. As the Arctic and Antarctic waters continue to warm, new species of fish are moving in at the expense of native fishes and other marine organisms. The impact of these changes in fish populations will not be well understood for some time.

Climate change is more than just a warming of the atmosphere. It is also a shift the patterns of weather. Some places will become wetter while others will become drier. Entire ecosystems could potentially be lost or shift in location. For example, warmer temperatures are leading to longer growing seasons and a melting of the permafrost, allowing trees to grow further northward at an unprecedented rate, shrinking the Arctic tundra. To complicate matters, the northern boreal forest may be expanding northward, but its southern limits may be reduced by increased forest fires resulting from warmer temperatures.

In the western United States, the snow pack is expected to decrease by 50% over the next 50 years. The reason for the loss of the snow pack is not necessarily a loss of precipitation, but warmer temperatures that will melt the snow earlier in the season. Rapid spring melting of the snow pack could cause several problems for the region. For example, less water will seep into the ground, potentially stressing the trees in the summer as temperatures warm and water is lost through transpiration. Second, mountain streams normally have higher water flow in the spring from snow melt, but maintain constant flows throughout the summer because snow melt in the mountains recharges the groundwater. The loss of winter snow pack could significantly alter the flows of western rivers where the rapid spring snow melts would cause very high flows followed by very low flows, or even drying of the river bed in the late summer.

On land, some species of animals will be able to track the changes in temperature and adjust their distribution accordingly. This is the climatic envelope hypothesis, which basically means that a species is restricted to a particular climate, similar to their fundamental niche. When the climate changes, their fundamental niche changes and their populations change in response to the new climate. We have already seen changes in species distribution for marine fish, butterflies, birds, and plants as they track their climate envelope.

However, not every species will be able to easily move to new regions. One good example are cold water trout who cannot survive temperatures much above 77°F (25°C). As temperatures warm in the summer above this critical temperature, these fishes may be unable to migrate to cooler waters and will perish. A criticism of the climate envelope hypothesis is that it only considers a single species. However, a recurring theme of life is that no species is an island, isolated from other species. If a predator is dependent on a particular prey item and that prey cannot track its climate envelope, then the predator must adapt to a new food source, or perish. In another scenario, imagine a flower that has evolved with a specific pollinator; if the flower cannot survive under a new climate, nor track its climate envelope, it will perish taking the pollinator with it.

In summary, life is already responding to current climate change. Many species will be able to simply move to suitable habitats while others will perish, becoming extinct. Entire ecosystems will be altered as new species move in and others are lost. Predicting these changes is incredibly difficult due to the complexity of these ecosystems. Based on what has happened to life during previous episodes of climate change, we will almost certainly experience another, or 6th mass extinction event in the next few hundred years. The one thing that separates this climate change event and the ensuing mass extinction is the presence of humans and how we have greatly altered the Earth, the topic of the final chapter.

Chapter 14

The Human Impacts to Life: How we are changing the world

Destroying rain forest for economic gain
Is like burning a renaissance painting to cook a meal
E.O. Wilson

The last word in ignorance is the man who says
of an animal or plant,
What good is it?
Aldo Leopold

Introduction

For nearly 4 billion years, life has existed on the Earth evolving, diversifying, and changing the Earth. Over time, new species evolve to take advantage of new environments, while others go extinct because they were outcompeted or cannot adapt to the new environment. Mass extinctions caused by climate change or meteor impacts have occurred at least 5 times in the past. After each event, life once again diversified to fill the empty niches. Every new species to evolve can potentially alter the environment, and so it is with a relative new comer to the world, a bipedal ape with an unusually large brain and an uncanny ability to alter the environment.

Conservation biology is the scientific study of biodiversity with the goal of protecting species and the ecosystems they reside in. It's a difficult field for several reasons. First, understanding how all the different species interact with each other in the ecosystem can be quite difficult and may take years to understand. Second, scientific studies often become quickly politicized because of their economic and cultural impacts. And lastly, not everyone places the same value on wilderness areas and diversity; unfortunately, this view is largely out of ignorance. As our world has become increasingly urbanized there has been a growing disconnect with nature as we preoccupy ourselves with our daily lives, social media, and content-free news.

Human activities today are causing the 6th mass extinction. Although it may be decades before we fully realize the extent of this extinction, for it is just really starting to get underway. I begin this chapter exploring the causes and consequences of the previous mass extinctions. The majority of this chapter is a brief introduction explaining how our activities are degrading ecosystems leading to extinctions. The last section is my opinion on the need for conservation, the future of humanity, and life on the Earth.

A Brief History of Mass Extinctions

The Earth may be 4.6 billion years old, but the last 0.6 billion years has witnessed the rapid rise of complex life and increasing diversity. It may have taken 4 billion years, but once animal life appeared in the oceans, it rapidly diversified into cnidarians, mollusks, arthropods, echinoderms and chordates, along with a lot of different types of worm-like animals. Once plants, fungus, and animals colonized the land, diversity continued to increase, in turn, creating more complex ecosystems. However, over the past 542 million years there have been some pitfalls to the slow, but continual rise in diversity caused by mass extinctions.

The first well known mass extinction occurred at the end of the Ordovician Period 444 million years ago as the world entered into an intense ice age. The lower temperatures and glaciers lowered the oceans level, reducing shallow water habitats, causing the extinction of numerous marine organisms, especially the mollusks. But, life rebounded over the next 70 million years and diversified as plants colonized the land eventually evolving to grow tall and forming the forest forests. Finally, the land was becoming green. Animals followed the plants, and they too colonized the land and diversified.

About 365 million years ago marked the end of the Devonian Period with the Earth's second mass extinction. So far, the reasons for this mass extinction are unclear. Although there is evidence that a period of increased volcanism warmed the climate and acidified the oceans. At the same time, several continents were beginning to converge together allowing species that were once isolated to mix for the first time. As a result, the spread of better adapted cosmopolitan species may have been a factor in this second mass extinction. A similar, but smaller extinction took place when South and North America were joined for the first time about 5 million years ago when the Isthmus of Panama formed. Many North America species migrated southward driving the extinction of South American species that had been isolated for millions of years.

The End Permian extinction 251 million years ago marked the largest mass extinction in the history of the Earth when an estimated 85-95% of all species went extinct in a time period of a few hundred thousand years. It brought about the end of the Paleozoic Era and ushered in the Mesozoic Era. This extinction was likely caused by a period of intense volcanism creating the Siberian traps, which holds 720,000 cubic miles of volcanic rock. That would be enough to cover 23% of the lower 48 in rock one mile deep! The intense volcanism drastically changed the climate and acidified the oceans. It took about 20 million years for diversity to fully recover from this mass extinction. Ironically, prior to the extinction, the ancestors to modern mammals were the dominant terrestrial vertebrates. But after the extinction, a lineage of reptiles evolved into the dinosaurs to become the dominant vertebrate on land.

The Mesozoic Era suffered a mass extinction about 201 million years ago possibly due to another period of intense volcanism. This fourth mass extinction didn't wipe out the dinosaurs, but it did cause significant losses to marine life. The fifth and last mass extinction occurred about 65.5 million years ago and was the second worse extinction in Earth's history. This one is perhaps the most famous for whiping out the dinosaurs after a successful 170-million-year reign. For several decades, the leading hypothesis to explain the loss of the dinosaurs has been a large meteor impact that devastated the Earth's ecosystems. However, a few million years prior to the meteor impact, the dinosaurs were rapidly declining, quite possibly due to another period of intense volcanism. This time, in a region of modern day India, known as the Deccan traps. The extinction of the dinosaurs may have been caused by this one-two punch from rapid climate change brought about by volcanism and a meteor impact.

From these "Big Five" mass extinctions, there are several take home messages. First, rapid climate change was almost certainly a cause for several of them, including the most severe. Second, diversity does recover after mass extinctions, but it takes millions of years to do so. Lastly, previous mass extinctions have paved the way for the evolution of new species millions of years later. The best example has been the rise of the mammals after the extinction of the dinosaurs.

The 6th Mass Extinction

Today, we are causing the 6th mass extinction through our activities. In just the last 150 years, three times as many birds and mammals have gone extinct than in the previous 200 years. Throughout time, there has always been a background extinction rate, which is about 1 species per 100 years per 10,000 species. That means if you live in an area with 10,000 species, about one would go extinct every hundred years, but the loss would be offset by the evolution of new species. The world today has an estimated 8.7 million species, so we would expect about 8-9 species to go extinct worldwide each year with natural extinction rate.

However, the current extinction rate is closer to 50 species per 100 years per 10,000 species. That means we are losing an estimated 400 species each year to extinction. The loss of species of plants and animals is so fast that scientists are unable to fully describe them before they go extinct, let alone learn anything about their lives or potential benefits. Some conservation biologists estimate that we will lose 65% of all species by 2100, or 5.6 million species in 100 years, which is about 56,000 species per year! The extinction rate has probably not reached such a high rate yet, but it is not an impossibility in the near future based on the impacts humans are having on the planet.

It can be hard to determine when a species is truly extinct, many are rare or elusive, or just plain hard to find. Most scientists do not like to declare a species extinct, even if they have not been able to find it after a great deal of effort. Nowhere is this more apparent than the amphibian declines in the last few decades in Central and South America. A fungus combined with habitat loss, pollution, and changing climates has caused severe population declines in numerous species of amphibians. Estimates based on their current global population declines indicate that their current extinction rate is 25,000 times higher than the natural background extinction rate for amphibians.

Other vertebrates are also suffering losses, one out five species of fish are at risk of extinction and nearly 50% of all mammal species are declining. About 14% of all birds are at risk, but birds may have already suffered through a major round of extinction about 2,000 years ago. By bringing livestock and rats, combined with over exploitation, it's been estimated that the spread of people across the Pacific may have caused the extinction of nearly 2,000 endemic bird species on isolated Pacific islands.

We stand to lose a lot, so the question becomes, HOW? How are humans causing one of the largest mass extinctions in the history of the planet? Unfortunately, there are lots of reasons as to how we are causing it, but it really comes down to a few critical problems that the world will have to eventually address. The biggest problem is the rapid population explosion of humans based on cheap energy from fossil fuels. Human populations are rapidly expanding and using resources at an unsustainable rate leading to rapid climate change, habitat loss, a spread of invasive species, overexploitation of resources, and pollution. And in many cases, these factors are confounding, further exacerbating population declines of wildlife and extinction rates.

The Human Population Explosion

To say that humans are like no other animal the world has ever seen could still be an understatement. Every species is unique; representing an unbroken lineage going back to the origin of life. We are a mammal and evolved by natural selection just like every other animal on the planet. But, humans are very unique in many ways. Unlike all other animals that have ever preceded us, we are the smartest in that we have the ability to alter our surroundings to our liking. We no longer have to evolve to match our environment.

All other species must reside within their fundamental niche and most are further restricted to their realized niche by competition or other biological factors. This is not the case for humans, if it's too cold for us, we put on warmer clothes, build a fire, turn up the heater, or drive to southern Florida in a giant RV for the winter. Likewise, if it becomes too hot, we wear lighter clothing, move indoors, turn on the AC, or drive back north in a giant RV for the summer. Humans are not restricted to living in a single habitat, in fact we live on all continents, and in every ecosystem. We are not dependent on a single source of food, but can eat just about anything.

Since the end of the last ice age over 10,000 years ago, the population of humans has continually grown and spread around the planet. Much of the growth can be attributed to a period of stable climate and the agricultural revolution, which freed people from the vagaries of unpredictable climates and a hunting and gathering lifestyle. It offered people more stability and opportunities to take up other professions. From this, civilization was born. It took nearly 200,000 years of human existence for the population to reach 1 billion in 1804, 123 years later it doubled again to 2 billion in 1927. The baby boomers born after World War II between 1945 and 1964 were the first generation of people born to experience a doubling of the population in a mere 33 years between 1960 and 1999, when the population reached 6 billion people. As of 2016, the world population has reached 7.4 billion people and continues to grow at a rapid rate. Fortunately, there are signs that the world's population growth rate is beginning to slow.

One factor above all else is largely responsible for the rapid rise in human populations, a cheap source of energy found in coal and oil. Although, humans are unique in our intellectual abilities to exploit or even create novel resources, nevertheless human populations have responded to a rapid influx of energy just like any other animal. The ability to cheaply grow and distribute food because of cheap fossil fuels has temporarily expanded the Earth's carrying capacity for humans. Today, we are burning through fossil fuels at an alarmingly rapid rate, returning carbon dioxide to the atmosphere that has been sequestered for hundreds of millions of years.

Our amazing rise as successful species has come at a price that we will pay. Eventually, we will run out of fossil fuels and unless we replace it with another cheap energy source, our population will decline to match the carrying capacity of our Earth, there is no way around this. Secondly, we are drastically altering our climate with unknown consequences, and lastly we are having a devastating impact on the world's diversity as we cut down forest to grow food and make room for our expanding population. Through our activities, we are directly causing a mass extinction, and unlike any other mass extinction, this is the first one caused directly by the actions and success of a single species.

Global Climate Change

The global climate is changing, rapidly, due almost entirely to the burning of fossil fuels returning carbon dioxide to atmosphere that has been sequestered for hundreds of millions of years. Climate change is a scientific fact backed up by repeated observations and experimental verification. Carbon dioxide levels have been rapidly rising in the atmosphere from the burning of fossil fuels and there is no other explanation to account for this change in the last century.

Climatologists have reconstructed the Earth's climate going back thousands of years and in some cases, hundreds of thousands of years. Regardless of the data used, whether it's from ice cores, tree rings, lake and marine sediments, or direct measurements, it all paints the same picture: *The climate is rapidly changing.*

Rapid climate change has been implicated in some of the largest mass extinctions of the Phanerozoic Eon (542 Million years ago – present), including the End Permian Extinction that wiped out 80-95% of all life on the Earth. Species go extinct during times of rapid change because they cannot evolve quickly enough to survive in the new environment or they are unable to migrate to new regions that are suitable to their physiological tolerances. For example, if you require wet and humid environment and it quickly dries from loss of precipitation; then you must evolve to survive in the drier environment, or migrate to wetter regions. If a species fails to adapt or move to suitable habitat, it will go extinct.

WHAT DOES LIFE DO?

Today, there is an estimated 8 million species, give or take a million. Some will be able to migrate to new habitats as the climate alters their habitats, but many will not. Additionally, the rate of climate change is so fast that many species will not be able to evolve to survive in their current locations and will face extinction. Global climate change may be one of the biggest factors leading to the loss of diversity worldwide.

Habitat Loss and Degradation

Imagine a priceless painting, *Starry Night* by Van Gogh. Intact it's priceless. Now imagine you cut it into small pieces. Every time a part is cut and removed the value of the painting is reduced, the small pieces by themselves are practically worthless. The same goes for ecosystems; intact they are priceless, as they become fragmented with roads, urbanization, or clear-cutting, their value becomes increasingly diminished. Losing species to extinction is like removing the brush strokes, at first it's not very noticeable, but with every loss, the quality of the picture loses its meaning and worth. As ecosystems lose species, they become a shell of their former selves.

Forests cover about 31% of the land on our planet. Each year, about 46-58 thousand square miles are lost to clear cutting, logging, fires, and degradation from 225, loss of disturbance, or climate change. That's an area about the size of Alabama lost, much of that is in the tropics where species diversity is the highest. Clearing the Amazon rain forest in Brazil is one of the biggest areas of concern. The land is being cleared mostly for cattle ranching, and small scale clearing for subsistence agriculture is also becoming a major problem. When you realize that tree diversity can reach 300 species per square acre in some regions, it's easy to see how the loss of forests are driving the loss of species. There are many more species of animals compared to plants. To estimate insect diversity in the tropics, an entomologists fumigated a single tree in Panama and collected 50 new species of beetles that were previously unknown to science.

Road construction fragments the landscape and alters the dynamics of the forests and making it inhospitable for many species. Studies of forests patches in Panama and in the Amazon have shown that smaller patches of forest house fewer species over time. What this means is that 10 forest patches, 10 acres each will have fewer species than a single large patch of 100 acres. Some animals, including large predators, require large intact tracts of land to survive and aren't adapted to a fragmented landscape. You would think that birds could easily move between forest fragments to maintain their populations, but some birds won't fly short distances across fragmented habitats.

The North American grasslands once formed vast prairies with herds of buffalo grazing on the native grasses. Plant diversity remained high from frequent fires and grazing that prevented any one species from becoming dominant. Today, the buffalo are long gone, hunted to the brink of extinction and the land has been converted to croplands. In Texas, Oklahoma, and New Mexico improved technologies in oil and natural gas extraction have led to a rapid rise in oil pads throughout the region. As you can expect, plant and animal diversity, especially grassland birds, such as prairie chickens and Sage Grouse, have declined rapidly in the past few decades from the loss of habitat from farming and resource extraction.

Habitat loss isn't limited to clearcutting and habitat fragmentation. The loss of natural cycles can also degrade habitats leading to a loss of diversity. In the southeastern U.S. routine fire would spread through Longleaf pine habitats about once every 3-6 years. Fire was an integral part of these ecosystems, a single fire started from summertime lightning strikes could last for weeks. As the fire burned across the landscape it removed undergrowth and fallen logs preventing the buildup of fuel on the ground. The frequent fires maintained an open park-like forests promoting high rates of plant and animal diversity. In some locations, plant diversity reached a high of 50 species per square meter.

Beginning in the 1920s, fire was actively suppressed resulting in the encroachment of hardwoods, which further prevented the routine fires and reduced diversity. The loss of fire also coincided with increased habitat fragmentation, and as a result, the Longleaf pine forests of the southeast have been severely degraded over time either through clear cutting or succession to hardwoods. Species like the Red-cockaded Woodpecker became endangered because they are totally dependent on pristine Longleaf pine ecosystems with natural fire cycles and an open understory. These small woodpeckers are not the only ones that have suffered losses from ecosystem degradation, other species including Indigo Snakes and Gopher Tortoises also dependent on the pine habitat have been severely reduced in populations. Today, the Longleaf pine ecosystem is one of the most endangered ecosystems on the planet.

Habitat loss and degradation are not limited to the land. Our streams and rivers have also become severely impacted by dams and the lakes they create. Most rivers naturally vary in flow throughout the year; a type of natural disturbance creating habitats in rivers and maintain connectivity to riparian areas. Throughout much of North America, high flows often occur predictably in the spring from snow melt. High flows can also occur less predictably when rivers swell due to increased rainfall. It is this natural flow variability, both predictable and random, that is important for maintaining riparian areas adjacent to the rivers and creating different types of habitats in the river itself. Dams fragment streams, reduce flow variability, confining rivers to their main channel and separating them from their adjacent riparian areas. The loss of spring flooding is similar to the loss of fire Longleaf pine forests, the change in disturbance regimes simplifies the ecosystems and results in a loss of diversity.

Many species of fish including trout, sturgeon, and eels rely on rivers being connected to the oceans so they can complete their life cycle. Eels and sturgeon must return to the oceans to reproduce, whereas salmon must return to rivers to reproduce. If dams impede their migrations, then those migratory species will become extirpated from those rivers. Additionally, some species of fish produce semi-buoyant eggs that must float down a river to suitable habitat where the fish hatch, grow, and then swim back upstream. Large dams produce lakes that are unsuitable to these pelagic spawning fish. They lay their eggs, but they float down into lakes where the young fish are unable to survive.

Perhaps one of the most devastating forms of habitat destruction that goes largely unseen is trawling on the continental shelf, especially for shrimp. Trawling involves dragging nets across the bottom of the ocean behind a large boat. These nets indiscriminately catch everything in their path, while destroying the marine habitat in the process. In 2001, satellite trackers were used to monitor and map the movement of boats trawling for shrimp in coastal waters. It was discovered that they trawl an area about the size of the entire continental shelf every 18 months. In some places it can take decades for the area to fully recover from a single trawling incidence. Unfortunately, the habitat loss goes largely unnoticed by everyone because it is underwater and not readily observed unlike deforestation that can easily be seen or tracked with satellite imagery.

Invasive Species

When a species is introduced into a new habitat, becomes established and spreads to new areas while displacing native species, then it is called an invasive species. When a species becomes invasive, it can easily displace the native species disrupting entire ecosystems. Common examples of invasive species you encounter almost every day includes pigeons, House (English) Sparrows, and European Starlings, all brought over from Europe. There are numerous reasons why a species can become invasive. Often, invasive species are habitat generalist and they are taking advantage of degraded ecosystems. Sometimes it's because they escaped their natural enemies allowing their populations to grow and expand unchecked. Or the invasive species is a superior competitor compared to the native species.

In the Eastern United States, the aquatic plant *Hydrilla* has spread rapidly in both lakes and rivers. As it spreads, it displaces the native vegetation and alters fish communities. Its thick growth limits fish populations and skews them to smaller individuals. Bird populations dependent on native fish and snails can also be harmed by the presence of *Hydrilla*, as it completely degrades the habitat for most native organisms.

The deciduous forests of Eastern North America have suffered through several major tree die-offs in the last hundred years. In the early 1900s, a pathogenic fungus spread rapidly throughout the east wiping out native American Chestnut trees by the 1940s. Prior to the chestnut blight, the American Chestnut tree was the dominant hardwood of the region. More recently, the woolly adelgid, a small insect that was introduced to Virginia in the 1950s, has decimated the Eastern Hemlock tree. It feeds by sucking the sap of hemlock and spruce trees, eventually killing the tree. As of 2015, 90% of the geographic range of the hemlock trees is threatened by the woolly adelgid. Its spread northward has been limited by extreme cold winters, however recent warming from climate change may allow this invasive species to move northward, killing more trees.

Remote islands in the Pacific are often home to endemic birds found nowhere else in the world, and they also serve as nesting grounds for millions of sea birds. Having never been connected to mainland areas, these islands were free of many predators including rats and snakes. During World War II, cargo and war ships fighting in the Pacific accidently transported brown tree snakes to the Island of Guam. Within a few years the snake's population exploded due to lack of natural predators and abundant resources. The result of the introduction was that they wiped out almost all the endemic birds on Guam in a few decades.

Introduced rats are a major problem for nesting birds on oceanic islands. Mammals are unable to naturally colonize remote islands because they cannot survive drifting for days on the ocean, they die after a few days without food or water. Humans have spread rats to numerous oceanic islands where they have become established. Rats are omnivores, a food generalists capable of eating a variety of foods including plant and animal material. They can cause harm to plants by eating their seeds and fruits. Native bird populations on islands have evolved in the absence of natural predators, so they are mostly defenseless against rats that eat their eggs. For example, many seabirds nest on the ground, where the rats are known to eat the eggs and the young. To make matters worse, rats easily climb trees so they can eat bird eggs and nestlings in the trees and on the ground.

The eradication of invasive species can also become quite complicated and actually cause harm to the remaining native species. When invasive plants displace native plants, some native animals will form new relationships with the invasive species. In the desert southwest, the Willow Flycatcher is able to successfully nest in non-native salt cedars that have replaced much of the native vegetation in riparian areas. If managers attempt to remove the salt cedar too fast without replacing it with other vegetation, then its removal could negatively impact the Willow Flycatcher. To further complicate the problem, the loss of natural flows and prolonged drought has made conditions favorable for the salt cedar and less favorable for native vegetation, including cottonwoods

Overexploitation

Standing at the ocean's edge, its vast expanses seem to go on forever. To some, it would seem almost impossible to deplete its resources, yet that is exactly what we are doing. The detrimental changes taking place under the waves remain largely unseen and are due largely to overexploitation of its resources

Overfishing is a huge problem worldwide and is causing major changes to the marine ecosystems. Unfortunately, the issues that are causing overexploitation are complicated and not easily solved. Part of the reason for overfishing arises from the fact that fish are a great source of protein. In fact, salmon has actually been labeled a super food, or one of the healthiest foods in the world, so we should be eating more of it.

Because fish are a highly desirable food, economics becomes a major factor driving declines in fisheries. From a business point of view, if you have one fishing boat and turn a good profit, then it's in your economic interest to reinvest the money into more fishing boats to make more profits. Eventually, the fishing fleets grow larger, taking more and more fish from the oceans. Another problem causing the decline in fisheries from what is called the tragedy of the commons. If there is a common resource that is unregulated, then it is in your best interest to fully exploit the resource, or someone else will exploit it. Fish in the open ocean are a common resource available to everyone. Unfortunately, they are overexploited because the biggest immediate benefit goes to the one who most exploits the resource. The drive for short-term profits come at a costs of long-term sustainability.

Additional problems arise because it can be difficult to obtain accurate information on annual takes of a commercial fish, or to determine the size of their populations. Large-scale industrial fishing can collapse a fishery within ten years to one-tenth its original size. Most commercial fish populations can sustain a small harvest, but when too many fish are taken, their populations can quickly collapse as in the case sardines, anchovies, and the Atlantic Cod. In 1992, populations of Atlantic Cod dropped to 1% of their historical levels, driven mostly by improved equipment and technology.

The collapses in fish populations are especially troubling in long-lived species including sharks and other large apex predators. Studies have shown that more than 90% of larger fish have been removed from the oceans mostly due to long-lines, drift-nets, and other industrial fishing practices. Large tuna and sharks have been hit especially hard. Shark populations are collapsing world-wide, and some species have declined 97-99%. Perhaps the saddest part of the shark story is that most of them are caught only to have their fins removed to make shark-fin soup. The rest of the animal is often thrown overboard to die.

The impact from the loss of large apex predators on marine ecosystems is just now being fully understood. For example, parrotfish are found on coral reefs throughout the world where some species eat parts of the coral, creating sand and preventing algae growth. One of their natural predators are sharks. Unfortunately, the decline in shark populations has led to higher populations of parrotfish, which in turn, eat more corals. As you can see, the loss of sharks can lead to further degradation of coral reefs by allowing one species to become overabundant, directly causing harm to the reef.

Trawling for shrimp, a problem I discussed in habitat degradation, is also a problem of overexploitation. Not only does trawling destroy the habitat of the target species, including shrimp, it also catches every species in the area. Marine sea turtles are especially hard hit from shrimp trawling as they get caught in the nets and drown. Sea turtles also face problems on their nesting grounds as local people often collect and eat the eggs, further exacerbating their population declines. As a result, all sea turtles in the US waters are listed as endangered species.

On land, overexploitation is also a problem. In the 1800s, the Passenger Pigeon in North America may have been one of the most abundant birds in the history of the world with population in somewhere between 3.5 and 5 billion birds. They were wiped out by widespread commercial hunting and habitat loss. Between 1870 and 1890 their populations went into rapid decline and by the 1890s their populations had declined to the point where extinction was all but a certainty. The last Passenger Pigeon died in captivity in 1914. The extinction of the Passenger Pigeon serves as hard lesson that even very abundant species can be driven to extinction through rampant overexploitation and loss of habitat.

Along the Atlantic and Gulf of Mexico shores, horseshoe crabs crawl out of the water during spring high tides to lay their eggs at the water line as they have done for 450 million years. Thousands of migrating shorebirds time their migration to coincide with the laying of the horseshoe crab eggs. With the collapse of the lobster industry (from over harvesting), fishermen have turned to eel fishing and collecting horseshoe crabs for bait. At its height in the late 1990s, over 2.5 million horseshoe crabs were collected. As you can guess, their populations rapidly declined and so did the populations of shorebirds. This is another example illustrating that species are not isolated, they exist in communities where the loss of one species can directly lead to the loss of other species. The Red Knot, is a tragic example, whose populations have declined by over 70% in the last decade due to the loss of horseshoe crabs from overharvesting.

Overharvesting is not limited to catching animals for food or bait. Some organisms are being overharvested by collectors for the pet trade.

Many species of reptiles, including box turtles, snakes and lizards, and wild parrots are harvested for the pet trade. Unfortunately, there is very little data present to fully determine the impact of harvesting for the pet trade. Another major problems occurs in the aquarium trade where collectors harvest tropical marine fish. Some areas such as Australia and Hawaii closely regulate this type of fish harvest, keeping the industry sustainable. Unfortunately, it's not so well regulated in other regions where local people use dynamite or cyanide to "stun" the fish. The problems with exploding dynamite underwater are apparent when numerous fish and corals are immediately killed. The problems with cyanide are more subtle, most of the fish harvested with cyanide will usually die within a few months from liver failure and reefs where cyanide is used will begin to die. Both of these methods of fish harvest are unsustainable for the simple fact that the collecting methods destroy the habitat required for the fish they are harvesting.

Overharvesting is not limited to animals in the pet trade. Currently, populations of many cactus species in the southwest and Mexico are in sharp decline from unregulated collecting. They are sold to collectors mainly in the U.S. and Europe. Similar to cacti, orchids are overharvested by collectors as well.

The problems do not stop here; bears, rhinos, and sea horses are collected for traditional Asian medicines. In African countries suffering from war, famine, or poverty, people turn to the local mammals and hunt them for "bush meat", causing the bush meat crises where many large primates are being killed for food. Most recently, Elephants declined by 30% between 2007 and 2014 from illegal poaching for their tusks to be sold in Asian markets.

Pollution

We have become a disposable planet. The waste and byproducts of our civilization is staggering. The widespread use of plastics has become a major problem because they are cheap to make, but they take a long time to break down. Each year, about eight million tons or 16 billion pounds of plastic are dumped into the oceans each year. The plastic doesn't go away; instead plastic waste gets consumed by animals causing it to accumulate in the food chain. So far, at least 700 species of marine organisms have ingested plastic as part of their diet. Tragically, plastic bags resembling jellyfish are eaten by giant Leatherback sea turtles where it can make them sick or kill them directly. Seabirds and whales can also be killed by plastic debris when they accidently consume it. They aren't equipped to handle the plastic and end up choking to death on it.

Pollution from coal-fired plants releases nitrogen and sulfur compounds into the atmosphere where they react with water to form acid rain. In the eastern U.S., acid rain from coal-fired plants and automobiles has caused fish kills in some lakes and has acidified the ground so much in some places it has even killed trees. Coal also contains heavy metals including mercury, lead, arsenic, and cadmium. Modern coal-fired electric plants remove much of the waste through filtration, but these waste have to be dealt with so they don't leach heavy metals into the ground or into the drinking water supply.

The mining of coal also poses numerous problems and West Virginia serves as prime example. In order to extract the coal, which is often buried deep underground, large areas of mountains are removed or large pits are dug. A problem with mining Is that the tailings from these large scale operations are often dumped into streams polluting them with heavy metals. Environmental degradation is not limited to coal mining. Acid mine drainage from mines that target gold, silver, copper, or other metals, is also a severe problem; especially throughout the west where the lower pH kills fish and invertebrates.

Untreated sewage is another major problem, especially in poor regions around the world. It is well known that drinking contaminated water leads to numerous health problems in people. But, it also causes problems for the rivers, lakes, and marine environments. Earlier, we learned that aquatic ecosystems are limited by nutrients. Untreated sewage entering our water ways serves as a major source of nutrients directly leading to large algae blooms. In aquatic systems, large algae blooms can quickly lead to eutrophic conditions capable of causing large fish kills. Fertilizer runoff from farms can also cause algal blooms and dead zones too. A striking example of a dead zone occurs in the Gulf of Mexico where the Mississippi River brings tons of nutrients to region.

To the Future

For every complex problem, there is almost always a simple answer that is wrong. No matter what we do, humans will impact the world, nowhere is this truer than in the conservation of diversity. The most pressing problem is the rampant growth of the human population, which is causing the problems I previously discussed. Unfortunately, our economic models are based on continuous population growth. Every month a job reports comes out reporting the addition (or loss) of new jobs. A growing economy is great because it generates wealth and improves our standard of living. Unfortunately, the population cannot keep growing indefinitely, there are just not enough resources on this planet to sustain an ever increasing population. And most people are vehemently against birth restrictions. However, most economists agree that conservation is better for the economy in terms of job growth and higher earnings in the long run. A commercial fisherman may lose his job as a fisherman, but he may make more money in ecotourism by repurposing his boat from fishing to accommodating divers. A coal miner may earn the same amount of money manufacturing solar panels, but will be healthier from not having to work in harsh conditions.

As the human population continues to grow and the world becomes increasingly connected, biodiversity will continue to decline. Basically, humans are simplifying the Earth's ecosystems through urbanization, clear cutting, habitat fragmentation, disrupting natural disturbance regimes, and spreading invasive species. All of these factors are contributing to the world-wide decline in biodiversity as it becomes increasingly homogenized. As ecosystems become increasingly fragmented and degraded, fewer native species will remain. Eventually, specialized species dependent on unique habitats will be replaced by a few generalist species capable of cohabitating with humans or living in disturbed habitats.

Climate change is exacerbating the decline in biodiversity caused by habitat loss and the spread of invasive species. Native species must exist within their fundamental niche; as the climate changes they must be able to move to and find suitable habitat. Unfortunately, many species will have a very difficult time moving to suitable habitat if the climate changes because a once continuous habitat has been chopped into many smaller fragments. And, current nature preserves may have their ecosystems altered as a result of climate change and the spread of invasive species, making them unsuitable for the species they were meant to preserve.

Importantly, climate change will also affect our civilization. As a species we are young, about 200,000 years old, civilization is less than 10,000 years old. The rise of civilization was brought about by the agricultural revolution, coinciding with a period of very stable climate. By having a stable climate, our ancestors could predict when to plant and harvest their crops. Global climate change threatens the stability of our climate, which could make it hard to grow the crops we need to feed everyone. A 20-foot rise in sea levels would displace hundreds of millions of people from coastal regions. States like Florida will become inundated causing major economic losses for the region. Intensification of weather patterns could bring longer prolonged drought as we have seen in California, or an increase in severe thunderstorms, or hurricanes.

When it comes to climate change, less industrialized nations wish to improve their economies by utilizing coal and oil because they are cheap sources of energy. However, the long term health of their people will suffer by creating more pollution, also leading to further loss of diversity and ecosystem functioning. Although, switching to renewable resources would be more beneficial in the long run as less pollution would be generated leading to fewer environmental problems.

We are clearly causing the 6th mass extinction. Time will tell the extent of the loss of diversity. But, life will go on, humans will most likely continue to exists for some time on this planet; although most likely not with 7-8 billion people. It will take millions of years for biodiversity to recover to its previous levels before the arrival of humans. Perhaps we are entering a new era, we could call it the Homogenozoic Era.

The end of the Paleozoic Era ended with a mass extinction that wiped out 90% of all life, the Mesozoic Era ended with the extinction of the dinosaurs and nearly 80% of species. At our current rate, we could potentially match those losses. But this time the extinction is different, it was caused by the presence of another species, *Homo sapiens*. The name Homogenozoic comes from the fact that we are not only causing a mass extinction, we have moved species all over the planet, which will have a lasting effect on the future diversity of the planet.

The first period of the new era could be called the Anthropocene, named after the effects of man that caused one of the largest changes to life on the Earth in more than a quarter of a billion years, or perhaps even more! Who knows what future scientists will say about the time period we live in now. Hopefully, as a society we will make wise choices that will prevent the worst case scenarios of climate change and losses of diversity.

Life has come a long way since its ancient origins. It took over 3.8 billion years for the first truly sentient species to evolve. Our advanced technology has made life easier for us, greatly expanded our knowledge of the universe, and is also causing harm to our planet. Not to sound bleak, but much of the damage to our planet has already been done, and we are set on a course of rapid climate change and collapsing ecosystems leading to a great mass extinction. We can attempt to mitigate the severity of the problems by recycling, consuming less, conserving our resources, and voting for politicians that wish to serve our interest. But, one thing we have learned about life is that it has a tenacious ability to adapt and evolve. Ironically, the same cosmopolitan invasive species that we have spread around the Earth, will likely be the ancestors of future species. Our activities will leave a permanent mark on the Earth.

What about the fate of humanity? Well, here is where I really become speculative, so feel free to disagree with me. Maybe, I'll at least spark your imagination. My hypothesis is that humans, or at least our descendants, will exists on this planet for a very long time, tens of millions of years, perhaps until the sun ends its life more than 5 billion years in the future. Although, by then our future descendants may not be quite human as they will have been evolving for a long time. You may be wondering why I used the term hypothesis rather than opinion. Basically, it meets the requirements of a scientific hypothesis, it is testable and potentially falsifiable, although I will likely never know the answer. I base my hypothesis on observations of other long-lived species. First, we are widespread and found in a variety of habitats. Second, we are not dependent on a single resource or a limited number of resources for survival. Additionally, and perhaps most importantly, we no longer have to rely on evolution for us to adapt to a changing environment. Our unique intelligence allows us to quickly acclimate to a given situation. No other species has ever had this unique advantage. However, the continuation of our advanced civilization and high standard of living is totally dependent on cheap energy and creating a balance between population size and sustainable use of our resources.

www.ingramcontent.com/pod-product-compliance
Lightning Source LLC
Chambersburg PA
CBHW031115020426
42333CB00012B/99